Collins

GCSE 9-1
Maths Higher
in a week

Fiona Mapp

Revision Planner

Questions marked with the symbol 🚫 *should be attempted without using a calculator.*

Prime Factors, HCF and LCM

Prime Factors

Apart from **prime numbers**, any whole number greater than 1 can be written as a product of **prime factors**. This means the number is written using only prime numbers multiplied together.

A prime number has only two factors, 1 and itself. 1 is not a prime number.

The prime numbers up to 20 are:

2, 3, 5, 7, 11, 13, 17, 19

The diagram below shows the prime factors of 60.

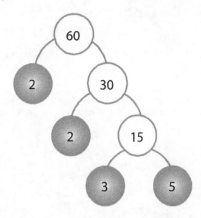

⬤ Divide 60 by its first prime factor, 2.

⬤ Divide 30 by its first prime factor, 2.

⬤ Divide 15 by its first prime factor, 3.

⬤ We can now stop because the number 5 is prime.

As a product of its prime factors, 60 may be written as:

$$60 = 2 \times 2 \times 3 \times 5$$

or in index form

$$60 = 2^2 \times 3 \times 5$$

Highest Common Factor (HCF)

The highest factor that two numbers have in common is called the **HCF**.

Example

Find the HCF of 60 and 96.

⬤ Write the numbers as products of their prime factors.

$$60 = 2 \times 2 \qquad \times 3 \times 5$$

$$96 = 2 \times 2 \times 2 \times 2 \times 2 \times 3$$

⬤ Ring the factors that are common.

$$60 = 2 \times 2 \qquad \times 3 \times 5$$

$$96 = 2 \times 2 \times 2 \times 2 \times 2 \times 3$$

⬤ These give the HCF $= 2 \times 2 \times 3$

$$= \mathbf{12}$$

Lowest (Least) Common Multiple (LCM)

The **LCM** is the lowest number that is a multiple of two numbers.

Example

Find the LCM of 60 and 96.

⬤ Write the numbers as products of their prime factors.

$60 = 2 \times 2 \qquad\quad \times 3 \times 5$

$96 = 2 \times 2 \times 2 \times 2 \times 2 \times 3$

⬤ 60 and 96 have a common factor of $2 \times 2 \times 3$, so it is only counted once.

$60 = \boxed{2} \times \boxed{2} \qquad\quad \times \boxed{3} \times 5$

$96 = \boxed{2} \times \boxed{2} \times 2 \times 2 \times 2 \times \boxed{3}$

⬤ The LCM of 60 and 96 is

$2 \times 2 \times 2 \times 2 \times 2 \times 3 \times 5$

$= \mathbf{480}$

SUMMARY

⬤ Any whole number greater than 1 can be written as a product of its prime factors, apart from prime numbers themselves (1 is not prime).

⬤ The highest factor that two numbers have in common is called the highest common factor (HCF).

⬤ The lowest number that is a multiple of two numbers is called the lowest (least) common multiple (LCM).

QUESTIONS

QUICK TEST

1. Write these numbers as products of their prime factors:

   a. 50   **b.** 360   **c.** 16

2. Decide whether these statements are true or false:

   a. The HCF of 20 and 40 is 4.

   b. The LCM of 6 and 8 is 24.

   c. The HCF of 84 and 360 is 12.

   d. The LCM of 24 and 60 is 180.

EXAM PRACTICE

1. Find the highest common factor of 120 and 42.
   [3 marks]

2. 🚫 Buses to St Albans leave the bus station every 20 minutes. Buses to Hatfield leave the bus station every 14 minutes.

   A bus to St Albans and a bus to Hatfield both leave the bus station at 10 am. When will buses to both St Albans and Hatfield next leave the bus station at the same time? [3 marks]

Fractions and Recurring Decimals

A **fraction** is part of a whole number. The top number is the **numerator** and the bottom number is the **denominator**.

There are four rules of fractions.

Addition $+$

You need to change the fractions so that they have the same denominator.

Example

$$\frac{5}{9} + \frac{1}{7}$$

The lowest common denominator is 63, since both 9 and 7 go into 63.

$$= \frac{35}{63} + \frac{9}{63}$$

$$= \frac{44}{63}$$

Remember to add only the numerators and not the denominators.

Subtraction $-$

You need to change the fractions so that they have the same denominator.

Example

$$\frac{4}{5} - \frac{1}{3}$$

The lowest common denominator is 15.

$$= \frac{12}{15} - \frac{5}{15}$$

$$= \frac{7}{15}$$

Remember to subtract only the numerators and not the denominators.

Multiplication $\times$

Before starting, write out whole or mixed numbers as improper fractions (also known as top-heavy fractions).

Example

$$\frac{2}{7} \times \frac{4}{5}$$

Multiply the numerators together.

$$= \frac{2 \times 4}{7 \times 5}$$

Multiply the denominators together.

$$= \frac{8}{35}$$

Division $\div$

Before starting, write out whole or mixed numbers as improper fractions (also known as top-heavy fractions).

Example

$$2\frac{1}{3} \div 1\frac{2}{7}$$

$$= \frac{7}{3} \div \frac{9}{7}$$

Convert to top-heavy fractions.

$$= \frac{7}{3} \times \frac{7}{9}$$

Take the reciprocal of the second fraction and multiply both fractions.

$$= \frac{49}{27}$$

$$= 1\frac{22}{27}$$

Rewrite the fraction as a mixed number.

Fraction Problems

You may need to solve problems involving fractions.

Example

Charlotte's take-home pay is £930. She gives her mother $\frac{1}{3}$ of this and spends $\frac{1}{5}$ of the £930 on going out. What fraction of the £930 is left?

Give your answer as a fraction in its simplest form.

$$\frac{1}{3} + \frac{1}{5}$$

This is a simple addition of fractions question.

$$= \frac{5}{15} + \frac{3}{15}$$

Write the fractions with a common denominator.

$$= \frac{8}{15}$$

$$1 - \frac{8}{15}$$

The question asks for the fraction of the money that is left, so subtract $\frac{8}{15}$ from 1.

$$= \frac{7}{15}$$

The fraction is in its simplest form.

Changing Recurring Decimals to Fractions

Recurring decimals are **rational** numbers because we can change them to fractions.

Examples

1. Change $0.1\dot{5}$ to a fraction in its lowest terms.

   Let $\quad x = 0.151\,515\ldots$ ①

   then $100x = 15.151\,515\ldots$ ②

   > Multiply by 10^n, where n is the length of the recurring pattern.

   Subtract equation ① from equation ②.

   ② − ①　$99x = 15$

   $$x = \frac{15}{99}$$

   $$x = \frac{5}{33} \quad \boxed{\text{Check } 5 \div 33 = 0.\dot{1}\dot{5}}$$

2. Change $0.3\dot{7}$ into a fraction.

   $$x = 0.3777\ldots \text{ ①}$$

   $$10x = 3.7777\ldots \text{ ②}$$

   > Multiply by 10, so that the part that is not recurring is in the units position.

   $$100x = 37.7777\ldots \text{ ③}$$

   > Multiply equation ① by 100.

   Subtract equation ② from equation ③.

   ③ − ②　$90x = 34$

   $$x = \frac{34}{90}$$

   $$x = \frac{17}{45} \quad \boxed{\text{Always check to see if your fraction simplifies.}}$$

SUMMARY

- To add or subtract fractions, write them using the same denominator.

- To multiply fractions, multiply the numerators and multiply the denominators.

- To divide fractions, take the reciprocal of the second fraction and multiply the fractions together.

- When multiplying and dividing fractions, write out whole or mixed numbers as top-heavy fractions before you begin the calculation.

- Recurring decimals are rational numbers, so they can be changed to fractions.

QUESTIONS

QUICK TEST

1. Work out the following:

   a. $\frac{2}{3} + \frac{1}{5}$　　　　b. $2\frac{6}{7} - \frac{1}{3}$

   c. $\frac{2}{9} \times \frac{5}{7}$　　　　d. $\frac{3}{11} \div \frac{22}{27}$

2. Change the following recurring decimals into fractions. Write each fraction in its simplest form.

   a. $0.\dot{7}$　　　b. $0.\dot{2}1\dot{5}$　　　c. $0.3\dot{5}$

EXAM PRACTICE

1. 🚫 In a magazine $\frac{3}{7}$ of the pages have advertisements on them.

   Given that 12 pages have advertisements on them, work out the number of pages in the magazine. [2 marks]

2. Prove that $0.\dot{3}\dot{6}$ is equivalent to $\frac{4}{11}$ [3 marks]

Indices

An **index** is sometimes called a **power**.

The base → a^b ← The index or power

Laws of Indices

The laws of indices can be used for numbers or algebra. The base has to be the same when the laws of indices are applied.

$$a^n \times a^m = a^{n+m}$$

$$a^n \div a^m = a^{n-m}$$

$$(a^n)^m = a^{n \times m}$$

$$a^0 = 1$$

$$a^1 = a$$

$$a^{-n} = \frac{1}{a^n}$$

$$a^{\frac{1}{m}} = \sqrt[m]{a}$$

$$a^{\frac{n}{m}} = (\sqrt[m]{a})^n$$

Examples with Numbers

1. Simplify the following, leaving your answers in index notation.

   a. $5^2 \times 5^3 = 5^{2+3} = \mathbf{5^5}$

   b. $8^{-5} \times 8^{12} = 8^{-5+12} = \mathbf{8^7}$

   c. $(2^3)^4 = 2^{3 \times 4} = \mathbf{2^{12}}$

2. Evaluate: Evaluate means to work out.

   a. $4^2 = 4 \times 4 = \mathbf{16}$

   b. $5^0 = \mathbf{1}$

   c. $3^{-2} = \frac{1}{3^2} = \mathbf{\frac{1}{9}}$

   d. $36^{\frac{1}{2}} = \sqrt{36} = \mathbf{6}$

   e. $8^{\frac{2}{3}} = (\sqrt[3]{8})^2 = 2^2 = \mathbf{4}$

3. Simplify the following, leaving your answers in index form.

   a. $7^2 \times 7^5 = \mathbf{7^7}$

   b. $6^9 \div 6^2 = \mathbf{6^7}$

   c. $\frac{3^7 \times 3^2}{3^{10}} = \frac{3^9}{3^{10}} = \mathbf{3^{-1}}$

   d. $7^9 \div 7^{-10} = \mathbf{7^{19}}$

4. Evaluate:

   a. $3^3 = 3 \times 3 \times 3 = \mathbf{27}$

   b. $7^0 = \mathbf{1}$

   c. $64^{\frac{1}{3}} = \sqrt[3]{64} = \mathbf{4}$

   d. $81^{\frac{1}{2}} = \sqrt{81} = \mathbf{9}$

   e. $5^{-2} = \frac{1}{5^2} = \mathbf{\frac{1}{25}}$

   f. $\left(\frac{4}{9}\right)^{-2} = \left(\frac{9}{4}\right)^2 = \frac{81}{16} = \mathbf{5\frac{1}{16}}$

Examples with Algebra

1. Simplify the following:

   a. $a^4 \times a^{-6} = a^{4-6} = a^{-2} = \dfrac{1}{a^2}$

   b. $5y^2 \times 3y^6 = \mathbf{15y^8}$

   | The numbers are multiplied. | The indices are added. |

   c. $(4x^3)^2 = \mathbf{16x^6}$

   Remember to square the 4 as well.

   If in doubt, write it out: $(4x^3)^2 = 4x^3 \times 4x^3$

   $$= \mathbf{16x^6}$$

   d. $(3x^4y^2)^3 = \mathbf{27x^{12}y^6}$

   or $3x^4y^2 \times 3x^4y^2 \times 3x^4y^2 = \mathbf{27x^{12}y^6}$

   e. $(2x)^{-3} = \dfrac{1}{(2x)^3} = \dfrac{1}{\mathbf{8x^3}}$

2. Simplify:

   a. $\dfrac{15b^4 \times 3b^7}{5b^2} = \dfrac{45b^{11}}{5b^2} = \mathbf{9b^9}$

   b. $\dfrac{16a^2b^4}{4ab^3} = \mathbf{4ab}$

3. Simplify:

   a. $7a^2 \times 3a^2b = \mathbf{21a^4b}$

   b. $\dfrac{14a^2b^4}{7ab} = \mathbf{2ab^3}$

   c. $\dfrac{9x^2y \times 2xy^3}{6xy} = \dfrac{18x^3y^4}{6xy}$

   $= \mathbf{3x^2y^3}$

QUESTIONS

QUICK TEST

1. Simplify the following, leaving your answers in index form.

   a. $6^3 \times 6^5$ **b.** $12^{10} \div 12^{-3}$

   c. $(5^2)^3$ **d.** $64^{\frac{2}{3}}$

2. Simplify the following:

   a. $2b^4 \times 3b^6$ **b.** $8b^{-12} \div 4b^4$

   c. $(3b^4)^2$ **d.** $(5x^2y^3)^{-2}$

EXAM PRACTICE

1. ⃠ Evaluate:

   a. 5^0 [1 mark]

   b. 7^{-2} [1 mark]

   c. $64^{\frac{1}{3}} \times 144^{\frac{1}{2}}$ [2 marks]

   d. $27^{-\frac{2}{3}}$ [2 marks]

2. ⃠ Simplify:

   a. $\dfrac{x^4 \times x^7}{x^{15}}$ [2 marks]

   b. $\dfrac{3x^4 \times 4x^2}{2x^3}$ [2 marks]

Standard Index Form

Standard index form (standard form) is useful for writing very large or very small numbers in a simpler way.

When written in standard form a number will be written as:

A number between 1 and 10 $1 \leqslant a < 10$	→ $a \times 10^n$

The value of n is the number of places the digits have to be moved to return the number to its original value.

If the number is 10 or more, n is positive.

If the number is less than 1, n is negative.

If the number is 1 or more but less than 10, n is zero.

Examples

1. Write 2 730 000 in standard form.

 ● 2.73 is the number between 1 and 10 $(1 \leqslant 2.73 < 10)$

 ● Count how many spaces the digits have to move to restore the original number.

 The digits have moved 6 places to the left because it has been multiplied by 10^6

 2.7 3

 2 7 3 0 0 0 0

 So, 2 730 000 = **2.73 × 10⁶**

2. Write 0.000 046 in standard form.

 ● Put the decimal point between the 4 and 6, so the number lies between 1 and 10.

 ● Move the digits five places to the right to restore the original number.

 ● The value of n is negative.

 So, 0.000 046 = **4.6 × 10⁻⁵**

On a Calculator

To put a number written in standard form into your calculator you use the following keys:

×10ˣ	EXP	or	EE

For example, $(2 \times 10^3) \times (6 \times 10^7) = 1.2 \times 10^{11}$ would be keyed in as:

| 2 | ×10ˣ | 3 | × | 6 | ×10ˣ | 7 | = |

or | 2 | EXP | 3 | × | 6 | EXP | 7 | = |

Doing Calculations

Examples

Work out the following using a calculator. Check that you get the answers given here.

1. $(6.7 \times 10^7)^3 = \mathbf{3.0 \times 10^{23}}$ (2 s.f.)

2. $\dfrac{(4 \times 10^9)}{(3 \times 10^4)^2} = \mathbf{4.\dot{4}}$

3. $\dfrac{(5.2 \times 10^6) \times (3 \times 10^7)}{(4.2 \times 10^5)^2} = \mathbf{884.4}$ (1 d.p.)

Examples

On a non-calculator paper you can use indices to help work out your answers.

1. $(2 \times 10^3) \times (6 \times 10^7)$

$= (2 \times 6) \times (10^3 \times 10^7)$

$= 12 \times 10^{3+7}$

$= 12 \times 10^{10}$

$= 1.2 \times 10^1 \times 10^{10}$

$= \mathbf{1.2 \times 10^{11}}$

2. $(6 \times 10^4) \div (3 \times 10^{-2})$

$= (6 \div 3) \times (10^4 \div 10^{-2})$

$= 2 \times 10^{4-(-2)}$

$= \mathbf{2 \times 10^6}$

3. $(3 \times 10^4)^2$

$= (3 \times 10^4) \times (3 \times 10^4)$

$= (3 \times 3) \times (10^4 \times 10^4)$

$= \mathbf{9 \times 10^8}$

You also need to be able to work out more complex calculations.

Example

The mass of Saturn is 5.7×10^{26} tonnes. The mass of the Earth is 6.1×10^{21} tonnes. How many times heavier is Saturn than the Earth? Give your answer in standard form, correct to 2 significant figures.

$$\frac{5.7 \times 10^{26}}{6.1 \times 10^{21}} = 93\,442.6$$

Now rewrite your answer in standard form.

Saturn is 9.3×10^4 times heavier than the Earth.

SUMMARY

- Numbers in standard form will be written as $a \times 10^n$.
- $1 \leqslant a < 10$
- n is positive when the original number is 10 or more.
- n is negative when the original number is less than 1.
- n is zero when the original number is 1 or more but less than 10.

QUESTIONS

QUICK TEST

1. Write in standard form:

   a. 64 000

   b. 0.000 46

2. 🚫 Work out the following. Leave in standard form.

   a. $(3 \times 10^4) \times (4 \times 10^6)$

   b. $(6 \times 10^{-5}) \div (3 \times 10^{-4})$

3. Work these out on a calculator:

   a. $(4.6 \times 10^{12}) \div (3.2 \times 10^{-6})$

   b. $(7.4 \times 10^9)^2$

EXAM PRACTICE

1. a. Write 40 000 000 in standard form. [1 mark]

   b. Write 6×10^{-5} as an ordinary number.
   [1 mark]

2. 🚫 The mass of an atom is 2×10^{-23} grams. What is the total mass of 7×10^{16} of these atoms?

   Give your answer in standard form. [3 marks]

Surds

Surds

A **rational number** is one that can be expressed in the form $\frac{a}{b}$, where a and b are integers and $b \neq 0$.

Rational numbers include $\frac{1}{3}$, $0.\dot{7}$, $\sqrt{16}$, $\sqrt[3]{8}$, etc.

Irrational numbers cannot be expressed as a fraction $\frac{a}{b}$.

Irrational numbers include π, π^2, $\sqrt{2}$, $\sqrt{7}$, etc.

> Roots, such as square roots or cube roots that are irrational, are also called **surds**.

Manipulating Surds

When working with surds there are several rules to learn:

1. $\sqrt{a} \times \sqrt{b} = \sqrt{ab}$

Example: $\sqrt{3} \times \sqrt{5} = \sqrt{15}$

2. $(\sqrt{b})^2 = \sqrt{b} \times \sqrt{b} = b$

Example: $(\sqrt{5})^2 = \sqrt{5} \times \sqrt{5} = 5$

3. $\frac{\sqrt{a}}{\sqrt{b}} = \sqrt{\frac{a}{b}}$

Example: $\frac{\sqrt{10}}{\sqrt{2}} = \sqrt{\frac{10}{2}} = \sqrt{5}$

4. $(a + \sqrt{b})^2$

$= (a + \sqrt{b})(a + \sqrt{b})$

$= a^2 + 2a\sqrt{b} + (\sqrt{b})^2$

$= a^2 + 2a\sqrt{b} + b$

5. $(a + \sqrt{b})(a - \sqrt{b})$

$= a^2 - a\sqrt{b} + a\sqrt{b} - (\sqrt{b})^2$

$= a^2 - b$

Examples

1. Simplify $\sqrt{75}$

$\sqrt{75} = \sqrt{25} \times \sqrt{3}$

$\sqrt{75} = \mathbf{5\sqrt{3}}$

> Look for the highest perfect square, i.e. 25.

2. Expand and simplify $(\sqrt{3} + 2)^2$

$(\sqrt{3} + 2)(\sqrt{3} + 2)$

$= \sqrt{9} + 2\sqrt{3} + 2\sqrt{3} + 4$

$= \mathbf{7 + 4\sqrt{3}}$

3. Work out $\frac{(2 - \sqrt{2})(4 + 3\sqrt{2})}{2}$

Leave your answer in surd form.

$\frac{(2 - \sqrt{2})(4 + 3\sqrt{2})}{2}$

$= \frac{8 + 6\sqrt{2} - 4\sqrt{2} - 3(\sqrt{2})^2}{2}$

$= \frac{8 + 2\sqrt{2} - 6}{2}$

$= \frac{2 + 2\sqrt{2}}{2}$

$= \frac{2(1 + \sqrt{2})}{2}$

$= \mathbf{1 + \sqrt{2}}$

Rationalising the Denominator

Sometimes a surd can appear on the bottom of the fraction. It is usual to rewrite the surd so that it appears as the numerator. This is called **rationalising** the denominator.

Examples

1. Rationalise $\dfrac{4}{\sqrt{7}}$

$= \dfrac{4}{\sqrt{7}} \times \dfrac{\sqrt{7}}{\sqrt{7}}$ — Multiply the numerator and denominator of the fraction by the surd. In this case $\sqrt{7}$.

$= \dfrac{4\sqrt{7}}{(\sqrt{7})^2}$

$= \dfrac{4\sqrt{7}}{7}$

2. Given that $\dfrac{4 - \sqrt{18}}{\sqrt{2}} = a + b\sqrt{2}$, where a and b are integers, find the value of a and the value of b.

$\dfrac{(4 - \sqrt{18})}{\sqrt{2}} \times \dfrac{\sqrt{2}}{\sqrt{2}}$ — First, rationalise the denominator.

$= \dfrac{4\sqrt{2} - \sqrt{36}}{(\sqrt{2})^2}$

$= \dfrac{4\sqrt{2} - 6}{2}$ — Now simplify.

$= 2\sqrt{2} - 3$

Hence, $a = -3$ and $b = 2$

3. Rationalise $\dfrac{1}{5 + \sqrt{2}}$

$\dfrac{1}{5 + \sqrt{2}} \times \dfrac{5 - \sqrt{2}}{5 - \sqrt{2}}$ — To rationalise $\dfrac{1}{5 + \sqrt{2}}$ you multiply both the numerator and the denominator by $5 - \sqrt{2}$ (this is the denominator with the sign between the two terms changed).

$= \dfrac{5 - \sqrt{2}}{5^2 + 5\sqrt{2} - 5\sqrt{2} - (\sqrt{2})^2}$

$= \dfrac{5 - \sqrt{2}}{5^2 - (\sqrt{2})^2}$

$= \dfrac{5 - \sqrt{2}}{25 - 2}$

$= \dfrac{5 - \sqrt{2}}{23}$

SUMMARY

- Surds are irrational numbers, which are any real numbers that cannot be expressed as a fraction $\dfrac{a}{b}$, where a and b are integers and $b \neq 0$.

- There are five rules when working with surds:

$\sqrt{a} \times \sqrt{b} = \sqrt{ab}$

$(\sqrt{b})^2 = \sqrt{b} \times \sqrt{b} = b$

$\dfrac{\sqrt{a}}{\sqrt{b}} = \sqrt{\dfrac{a}{b}}$

$(a + \sqrt{b})^2 = a^2 + 2a\sqrt{b} + b$

$(a + \sqrt{b})(a - \sqrt{b}) = a^2 - b$

QUESTIONS

QUICK TEST

1. Express the following in the form $a\sqrt{b}$, simplifying the answers.

    a. $\sqrt{24}$

    b. $\sqrt{200}$

    c. $\sqrt{48} + \sqrt{12}$

2. Molly works out $(2 - \sqrt{3})^2$. She says the answer is 1. Decide whether Molly is correct, giving a reason for your answer.

3. Decide whether this is correct:

    $\dfrac{1}{\sqrt{2}} = \dfrac{\sqrt{2}}{2}$

EXAM PRACTICE

1. Given that $\dfrac{5 - \sqrt{75}}{\sqrt{3}} = a + b\sqrt{3}$, find the value of a and the value of b. [4 marks]

2. Rationalise the denominator, simplifying the answer. [2 marks]

    $\dfrac{3}{\sqrt{6}}$

3. Work out $\dfrac{(5 + \sqrt{5})(2 - 2\sqrt{5})}{\sqrt{45}}$

    Give your answer in its simplest form. [3 marks]

Upper and Lower Bounds

Measurements

Measurements are never exact. They can only be expressed to a certain degree of accuracy.

When measurements are quoted to a given unit, say the nearest metre, there is a highest and lowest value they could be.

The highest value is called the **upper bound**.

The lowest value is called the **lower bound**.

Example

A length, l, is rounded to 6.2 cm to the nearest millimetre. This length would really lie between:

$$6.15 \leqslant l < 6.25$$

The real value can be as much as half the rounded unit below or above the rounded off value.

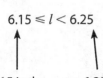

6.15 is the lower bound

6.25 is the upper bound

Finding Maximum and Minimum Possible Values of a Calculation

Some exam questions may ask you to calculate the upper or lower bounds of calculations.

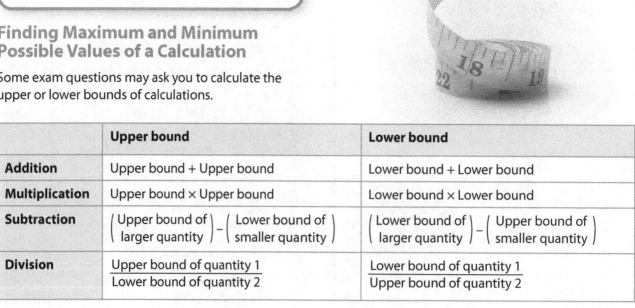

	Upper bound	Lower bound
Addition	Upper bound + Upper bound	Lower bound + Lower bound
Multiplication	Upper bound × Upper bound	Lower bound × Lower bound
Subtraction	$\left(\begin{array}{c}\text{Upper bound of}\\\text{larger quantity}\end{array}\right) - \left(\begin{array}{c}\text{Lower bound of}\\\text{smaller quantity}\end{array}\right)$	$\left(\begin{array}{c}\text{Lower bound of}\\\text{larger quantity}\end{array}\right) - \left(\begin{array}{c}\text{Upper bound of}\\\text{smaller quantity}\end{array}\right)$
Division	$\dfrac{\text{Upper bound of quantity 1}}{\text{Lower bound of quantity 2}}$	$\dfrac{\text{Lower bound of quantity 1}}{\text{Upper bound of quantity 2}}$

Examples

1. The side of a square measures 62 mm to the nearest mm. Work out the upper and lower bounds of the area of the square (in mm²).

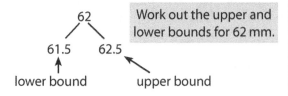

Work out the upper and lower bounds for 62 mm.

lower bound   upper bound

Upper bound of area = 62.5 × 62.5

$$= \mathbf{3906.25 \ mm^2}$$

Lower bound of area = 61.5 × 61.5

$$= \mathbf{3782.25 \ mm^2}$$

2. Given that $a = 4.2$ (to 1 d.p.) and $b = 6.23$ (to 3 s.f.), find the upper and lower bounds for the following calculations:

   a. $b - a$

   4.2
   4.15   4.25

   6.23
   6.225   6.235

   Upper bound = 6.235 − 4.15 = **2.085**

   Lower bound = 6.225 − 4.25 = **1.975**

   b. $\dfrac{b}{a}$

   Upper bound = $\dfrac{6.235}{4.15}$ = **1.5024**

   Lower bound = $\dfrac{6.225}{4.25}$ = **1.4647**

SUMMARY

- Measurements are expressed to a certain degree of accuracy.
- The highest value is called the upper bound.
- The lowest value is called the lower bound.

QUESTIONS

QUICK TEST

1. $c = \dfrac{(2.6)^3 \times 12.52}{3.2}$

   2.6 and 3.2 are correct to 1 decimal place. 12.52 is correct to 2 decimal places. Which of the following calculations gives the lower bound for c and which gives the upper bound for c?

   A $\dfrac{(2.65)^3 \times 12.525}{3.15}$

   B $\dfrac{(2.55)^3 \times 12.515}{3.15}$

   C $\dfrac{(2.65)^3 \times 12.525}{3.25}$

   D $\dfrac{(2.65)^3 \times 12.515}{3.15}$

   E $\dfrac{(2.55)^3 \times 12.515}{3.25}$

2. Work out the upper and lower bound of $\dfrac{6}{p}$, where p is 27 (rounded to the nearest whole number).

EXAM PRACTICE

1. The value of R is calculated by using this formula: $R = \dfrac{a - b}{b}$

   $a = 7.65$ correct to 2 decimal places.

   $b = 4.3$ correct to 1 decimal place.

   Find the difference between the lower bound of R and the upper bound of R. Give your answer to 3 significant figures.   [4 marks]

Formulae and Expressions

$a + b$ is an **expression**. $b = a + 6$ is a **formula**. The value of b depends on the value of a.

⬤ A term is a collection of numbers, letters and brackets, all multiplied together, e.g. $6a$, $2ab$, $3(x - 1)$.

⬤ Terms are separated by $+$ and $-$ signs. Each term has a $+$ or $-$ sign in front of it.

> $3ab$ means $3 \times a \times b$

> $3b^2$ means $3 \times b \times b$

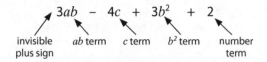

$$3ab \quad - \quad 4c \quad + \quad 3b^2 \quad + \quad 2$$

invisible plus sign ab term c term b^2 term   number term

Substituting into Formulae

Replacing a letter with a number is called **substitution**.

⬤ Write out the expression first and then replace the letters with the values given.

⬤ Work out the value, but take care with the order of operations, i.e. **BIDMAS**.

Examples

1. $a = 3b - 4c$

Find a if $b = 4$ and $c = -2$

$a = (3 \times 4) - (4 \times -2)$

$= 12 - (-8)$ ⬅ Subtracting a negative is the same as adding.

$= \mathbf{20}$

2. $E = \frac{1}{2}mv^2$

Find E if $m = 6$ and $v = 10$

$E = \frac{1}{2} \times 6 \times 10^2$

$E = \mathbf{300}$

Rearranging Formulae

The subject of a formula is the letter that appears on its own on one side of the formula.

Examples

1. Make a the subject of the formula $b = (a - 3)^2$

$$b = (a - 3)^2$$

$\pm \sqrt{b} = a - 3$ ⬅ Deal with the power first, square root both sides.

$\pm \sqrt{b} + 3 = a$ ⬅ Remove any term added or subtracted. Add 3 to both sides.

$$a = \pm \sqrt{b} + 3 \text{ i.e. } a = 3 \pm \sqrt{b}$$

2. Make x the subject of the formula $p = x^2 + y$

$$p = x^2 + y$$

$p - y = x^2$ ⬅ Subtract y from both sides.

$\pm \sqrt{p - y} = x$ ⬅ Square root both sides.

$$x = \pm \sqrt{p - y}$$

3. Make t the subject of the formula $v = u + at$

$$v = u + at$$

$v - u = at$ ⬅ Subtract u from both sides.

$\dfrac{v - u}{a} = t$ ⬅ Divide all of $v - u$ by a.

When rearranging a formula, sometimes the new subject appears in more than one term.

Example

Make x the subject of the formula $a = \dfrac{x+c}{x-d}$

$$a = \frac{x+c}{x-d}$$

$a(x-d) = x+c$ ← Multiply both sides by $(x-d)$.

$ax - ad = x+c$ ← Multiply out the brackets.

$ax - x = c + ad$ ← Collect like terms involving x on one side of the equation.

$x(a-1) = c+ad$ ← Factorise.

$$x = \frac{c+ad}{a-1}$$

These types of questions are usually worth at least 3 marks.

QUESTIONS

QUICK TEST

1. Simplify the following expressions:

   a. $6a - 3b + 2a - 4b$

   b. $3a^2 - 6b^2 - 2b^2 + a^2$

   c. $5xy - 3yx + 2xy^2$

2. If $a = \frac{3}{5}$ and $b = -2$, find the value of these expressions:

   a. $ab - 5$

   b. $a^2 + b^2$

   c. $3a - 6ab$

3. Make u the subject of the formula

   $v^2 = u^2 + 2as$

4. Make p the subject of the formula

   $q = \dfrac{p-t}{p+v}$

EXAM PRACTICE

1. 🚫 Sarah says, 'When $x = 2$ the value of $3x^2$ is 36'.

   Josh says, 'When $x = 2$ the value of $3x^2$ is 12'.

   Who is right? Explain why. [2 marks]

2. **a.** Work out the value of $y = 4(x-1)^2$ when $x = 4$. [2 marks]

   b. If $y = 4(x-1)^2$, make x the subject of the formula. [3 marks]

3. Make y the subject of the formula.

   $\dfrac{p-4y}{5y+q} = q$ [4 marks]

Brackets and Factorisation

Multiplying out brackets helps to simplify algebraic expressions.

Expanding Single Brackets

Each term outside the bracket is multiplied by each separate term inside the bracket.

$$5(x + 6) = 5x + 30$$

Examples

Expand and simplify:

1. $-2(2x + 4) = \mathbf{-4x - 8}$

2. $5(2x - 3) = \mathbf{10x - 15}$

3. $8(x + 3) + 2(x - 1)$ ← Multiply out the brackets.

   $= 8x + 24 + 2x - 2$ ← Collect like terms.

   $= \mathbf{10x + 22}$

4. $3(2x - 5) - 2(x - 3)$ ← Multiply out the brackets.

   $= 6x - 15 - 2x + 6$ ← Collect like terms.

   $= \mathbf{4x - 9}$

Expanding Two Brackets

Every term in the second bracket must be multiplied by every term in the first bracket.

Often, but not always, the two middle terms are like terms and can be collected together.

$$(x + 4)(x + 2) \quad = x^2 + 2x + 4x + 8$$
$$= x^2 + 6x + 8$$

Examples

Expand and simplify:

1. $(x + 4)(2x - 5) = 2x^2 - 5x + 8x - 20$

   $\qquad\qquad\quad = \mathbf{2x^2 + 3x - 20}$

2. $(2x + 1)^2 = (2x + 1)(2x + 1)$

   $\qquad\quad = 4x^2 + 2x + 2x + 1$

   $\qquad\quad = \mathbf{4x^2 + 4x + 1}$

> Remember that x^2 means x multiplied by itself.

3. $(3x - 1)(x - 2) = 3x^2 - 6x - x + 2$

   $\qquad\qquad\quad = \mathbf{3x^2 - 7x + 2}$

4. $(x - 4)(3x + 1) = 3x^2 + x - 12x - 4$

   $\qquad\qquad\quad = \mathbf{3x^2 - 11x - 4}$

5. $(2x + 3y)(x - 2y) = 2x^2 - 4xy + 3xy - 6y^2$

   $\qquad\qquad\qquad = \mathbf{2x^2 - xy - 6y^2}$

Expanding Three Brackets

When multiplying out three brackets, multiply out any two brackets to start with.

Example

Expand and simplify $(x + 3)(x + 5)(x - 2)$

$(x^2 + 8x + 15)(x - 2)$ ← Multiply out the first two brackets.

$(x^3 - 2x^2 + 8x^2 - 16x + 15x - 30)$ ← Multiply out the next brackets.

$= \mathbf{x^3 + 6x^2 - x - 30}$ ← Simplify.

Factorisation

Factorisation simply means putting an expression into brackets.

One Bracket

$4x + 6 = 2(2x + 3)$

To factorise $4x + 6$:

● Recognise that 2 is the HCF of 4 and 6.

● Take out the highest common factor.

● The expression is completed inside the bracket so that when multiplied out it is equivalent to $4x + 6$.

Two Brackets

Two brackets are obtained when a quadratic expression of the type $ax^2 + bx + c$ is factorised.

Examples

1. $x^2 + 4x + 3 = (x + 1)(x + 3)$

2. $x^2 - 7x + 12 = (x - 3)(x - 4)$

3. $x^2 + 3x - 10 = (x + 5)(x - 2)$

4. $x^2 - 64 = (x - 8)(x + 8)$ ◀

5. $81x^2 - 25y^2 = (9x - 5y)(9x + 5y)$

This is known as the 'difference of two squares'. In general, $x^2 - a^2 = (x - a)(x + a)$.

6. Here is a right-angled triangle.

   Four triangles are joined to enclose the square $MNPQ$. Work out the area of the square.

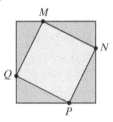

Area of the large square $= (a + b)^2$

Area of four triangles $= \dfrac{b \times h}{2} \times 4$

$$= \dfrac{a \times b}{2} \times 4 = 2ab$$

Area of square $MNPQ = (a + b)^2 - 2ab$

$$= (a + b)(a + b) - 2ab$$

$$= a^2 + 2ab + b^2 - 2ab$$

$$\mathbf{= a^2 + b^2}$$

SUMMARY

● To expand single brackets, multiply each term outside the bracket by everything inside the bracket.

● To expand two brackets, multiply each term in the second bracket by each term in the first bracket.

● Factorising means putting in brackets. Look for the highest common factor (HCF).

QUESTIONS

QUICK TEST

1. Expand and simplify:

   a. $(x + 3)(x - 2)$ **b.** $4x(x - 3)$

   c. $(x - 3)^2$

2. Factorise:

   a. $12xy - 6x^2$ **b.** $3a^2b + 6ab^2$

   c. $x^2 + 4x + 4$ **d.** $x^2 - 4x - 5$

   e. $x^2 - 100$

EXAM PRACTICE

1. Expand and simplify:

   a. $t(3t - 4)$ [1 mark]

   b. $4(2x - 1) - 2(x - 4)$ [2 marks]

2. Factorise:

   a. $y^2 + y$ [1 mark]

   b. $5p^2q - 10pq^2$ [2 marks]

   c. $(a + b)^2 + 4(a + b)$ [1 mark]

   d. $x^2 - 5x + 6$ [2 marks]

3. Show that $(a + b)^2 - 2b(a + b) = (a - b)(a + b)$ [3 marks]

Equations 1

Equations involve an unknown value that needs to be worked out.

Equations need to be kept balanced, so whatever is done to one side of the equation (for example, adding) also needs to be done to the other side.

Linear Equations of the Form $ax + b = c$

Examples

1. Solve: $3x = 15$

$x = \dfrac{15}{3}$ ← Divide both sides by 3.

$x = \mathbf{5}$

2. Solve: $\dfrac{x}{3} = 6$

$x = 6 \times 3$ ← Multiply both sides by 3.

$x = \mathbf{18}$

3. Solve: $5x - 2 = 13$

$5x = 13 + 2$ ← Add 2 to both sides.

$5x = 15$

$x = \dfrac{15}{5}$ ← Divide both sides by 5.

$x = \mathbf{3}$

4. Solve: $3x + 1 = 13$

$3x = 13 - 1$ ← Subtract 1 from both sides.

$3x = 12$

$x = \dfrac{12}{3}$ ← Divide both sides by 3.

$x = \mathbf{4}$

5. Solve: $\dfrac{x}{6} - 1 = 3$

$\dfrac{x}{6} = 3 + 1$ ← Add 1 to both sides.

$\dfrac{x}{6} = 4$

$x = 4 \times 6$ ← Multiply both sides by 6.

$x = \mathbf{24}$

Linear Equations of the Form $ax + b = cx + c$

Examples

1. Solve:

$7x - 4 = 3x + 8$

$7x = 3x + 12$ ← Add 4 to both sides.

$4x = 12$ ← Subtract $3x$ from both sides.

$x = \dfrac{12}{4}$

$x = \mathbf{3}$

Check by substituting 3 into both sides of the equation:

$7 \times 3 - 4 = 17$

$3 \times 3 + 8 = 17$

Since both the left-hand side of the equation and the right-hand side of the equation give the same answer, $x = 3$ is correct ✔

2. Solve:

$5x + 3 = 2x - 5$

$5x = 2x - 5 - 3$ ← Subtract 3 from both sides.

$5x = 2x - 8$

$5x - 2x = -8$ ← Subtract $2x$ from both sides.

$3x = -8$

$x = -\dfrac{8}{3}$

$x = \mathbf{-2\dfrac{2}{3}}$

Linear Equations with Brackets

Examples

1. Solve:

$$5(x - 1) = 3(x + 2)$$

$$5x - 5 = 3x + 6$$

$$5x = 3x + 11$$

$$2x = 11$$

$$x = \frac{11}{2}$$

$$x = \mathbf{5.5}$$

> Just multiply out the brackets, then solve as normal.

2. Solve:

$$5(2x + 3) = 2(x - 6)$$

$$10x + 15 = 2x - 12$$

$$10x = 2x - 12 - 15$$

$$10x = 2x - 27$$

$$10x - 2x = -27$$

$$8x = -27$$

$$x = -\frac{27}{8}$$

$$x = \mathbf{-3\frac{3}{8}}$$

3. Solve:

$$\frac{3(2x - 1)}{5} = 6$$

$$3(2x - 1) = 6 \times 5$$

> Multiply both sides by 5.

$$6x - 3 = 30$$

$$6x = 33$$

$$x = \frac{33}{6}$$

$$x = \mathbf{5.5}$$

Identities and Equations

- An **equation** can be solved to find an unknown quantity.
- An **identity** is true for all values of x. For example, $(2x + 1)^2$ is identically equal to $4x^2 + 4x + 1$. This can be written as:

$$(2x + 1)^2 \equiv 4x^2 + 4x + 1$$

QUESTIONS

QUICK TEST

Solve the following equations:

1. $2x - 6 = 10$ **2.** $5 - 3x = 20$

3. $4(2 - 2x) = 12$ **4.** $6x + 3 = 2x - 10$

5. $7x - 4 = 3x - 6$ **6.** $5(x + 1) = 3(2x - 4)$

EXAM PRACTICE

1. Solve the equations:

    a. $5x - 3 = 9$ [2 marks]

    b. $7x + 4 = 3x - 6$ [3 marks]

    c. $3(4y - 1) = 21$ [3 marks]

2. Solve:

    a. $5 - 2x = 3(x + 2)$ [3 marks]

    b. $\frac{3x - 1}{3} = 4 + 2x$ [3 marks]

3. Joe is asked to solve the equation $3(x - 6) = 42$

    Here is his working: $3(x - 6) = 42$

$$3x - 6 = 42$$

$$3x = 42 + 6$$

$$3x = 48$$

$$x = 16$$

What mistake did he make?   [1 mark]

Equations 2

Equation Problems

When solving equation problems, the first step is to write down the information that you know.

Example
The perimeter of this rectangle is 30 cm.

Work out the value of y and find the length of the rectangle.

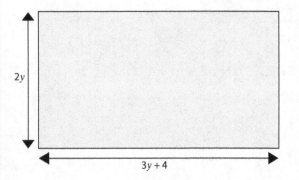

Write down what you know.

$$3y + 4 + 2y + 3y + 4 + 2y = 30$$

Simplify the expression and solve as normal.

$$10y + 8 = 30$$

$$10y = 30 - 8$$

$$10y = 22$$

$$y = 2.2$$

Length of rectangle $= 3 \times 2.2 + 4$

$$= \mathbf{10.6\,cm}$$

Solving Equations Involving Indices

Equations sometimes involve **indices**. You need to remember the laws of indices to be able to solve them.

Examples
Solve the following:

1. $y^k = \sqrt[3]{y} \div \dfrac{1}{y^5}$

$$y^k = y^{\frac{1}{3}} \div y^{-5}$$
— Rewrite as indices.

$$y^k = y^{\frac{1}{3} - (-5)}$$
— When dividing, subtract the indices.

$$y^k = y^{5\frac{1}{3}}$$
— Compare the indices since the bases are the same.

$$k = \mathbf{5\tfrac{1}{3}}$$

2. $\quad 2^{k+2} = 32$

$$2^{k+2} = 2^5$$
— Rewrite so that both bases are the same.

$$k + 2 = 5$$

$$k = \mathbf{3}$$

3. $3^k = 27$

$$3^k = 3^3$$

$$k = \mathbf{3}$$

Example

The area of this rectangle is 81 cm².

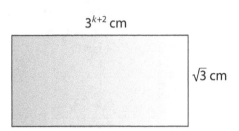

3^{k+2} cm

$\sqrt{3}$ cm

Work out the value of k.

$3^{k+2} \times \sqrt{3} = 81$ ← Write out the equation.

$3^{k+2} \times 3^{\frac{1}{2}} = 81$

$3^{(k+2+\frac{1}{2})} = 3^4$ ← Write out so that the bases are 3.

$3^{(k+\frac{5}{2})} = 3^4$ ← Add the indices of the left-hand side.

$k + \dfrac{5}{2} = 4$ ← Compare the indices.

$k = \dfrac{3}{2}$

QUICK TEST

1. The perimeter of this triangle is 60 cm. Work out the value of x and find the shortest length.

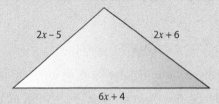

$2x - 5$ $2x + 6$

$6x + 4$

2. Solve $5^{k+3} = 5^{\frac{1}{3}}$

3. Solve $16^k = 64$

4. Solve $2^{3k-1} = 64$

EXAM PRACTICE

1. The sizes of the angles, in degrees, of the quadrilateral are:

$x + 30°$ $2x$ $x + 50°$ $x + 10°$

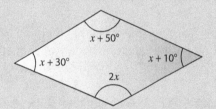

$x + 50°$

$x + 30°$ $x + 10°$

$2x$

Work out the smallest angle of the quadrilateral.
[3 marks]

2. The perimeter of the quadrilateral is three times the perimeter of the triangle. All measurements are in centimetres.

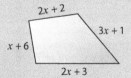

$2x + 2$

$3x + 1$

$x + 6$

$2x + 3$

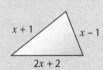

$x + 1$

$x - 1$

$2x + 2$

Work out the perimeter of the triangle. [4 marks]

3. Solve $16^{2k} = 64^{k+1}$ [4 marks]

SUMMARY

● **Equations involving indices are solved using the laws of indices.**

- **When multiplying, add the powers (provided the base is the same).**

- **When dividing, subtract the powers (provided the base is the same).**

- **When raising one power to another, multiply the powers.**

Quadratic and Cubic Equations

A **quadratic equation** can be written in the form $ax^2 + bx + c = 0$.

Quadratic equations can be solved using a variety of methods including **factorisation** or using the quadratic formula.

Method 1: Factorisation

● Write the equation in the form $ax^2 + bx + c = 0$

● Try to factorise the quadratic expression by putting it into brackets $(\quad)(\quad) = 0$

● Then solve the equation by making each bracket equal to 0.

Example

Solve $2x^2 - x - 3 = 0$

> Write out the two brackets and put an x and $2x$ in each one, since $x \times 2x = 2x^2$.

$(2x \quad)(x \quad)$

> We now need two numbers that multiply to give -3 (one positive and one negative) and when multiplied by the x terms add up to give $-1x$.

Try: $(2x + 1)(x - 3)$: Constant term $= 1 \times -3 = -3$ ✓

But: x terms $= -6x + x = -5x$ ✗ incorrect

Try: $(2x - 3)(x + 1)$: Constant term $= -3 \times 1 = -3$ ✓

x terms $= 2x - 3x = -x$ ✓ correct

> The $2x$ and 1 must be in different brackets.

A quick check gives
$(2x - 3)(x + 1) = 2x^2 - x - 3$

Now solve the equation:

$(2x - 3)(x + 1) = 0$

$\therefore (2x - 3) = 0$ so $x = \frac{3}{2}$

or $(x + 1) = 0$ so $x = -1$

$\therefore x = \frac{3}{2}$ **or −1**

Method 2: The Quadratic Formula

When a quadratic expression does not factorise, use the quadratic formula. For any quadratic equation written in the form $ax^2 + bx + c = 0$,

$$x = \frac{-b \pm \sqrt{b^2 - 4ac}}{2a}$$

Example

Solve the equation $2x^2 - 7x = 5$
Give your answers to 2 decimal places.

> Put the equation into the form $ax^2 + bx + c = 0$

$2x^2 - 7x - 5 = 0$

> Identify the values of a, b and c.

$a = 2, b = -7, c = -5$

> Substitute these values into the quadratic formula.

$$x = \frac{-b \pm \sqrt{b^2 - 4ac}}{2a}$$

$$x = \frac{7 \pm \sqrt{(-7)^2 - (4 \times 2 \times -5)}}{2 \times 2}$$

$$x = \frac{7 \pm \sqrt{49 - (-40)}}{4}$$

$$x = \frac{7 \pm \sqrt{89}}{4}$$

$$x = \frac{7 + \sqrt{89}}{4}$$
One solution is when we use $+ \sqrt{89}$
$x = $ **4.11** (2 d.p.)

$$x = \frac{7 - \sqrt{89}}{4}$$
One solution is when we use $- \sqrt{89}$
$x = $ **−0.61** (2 d.p.)

Check
$2 \times (4.11)^2 - 7(4.11) - 5$
$= 0$ ✓

Check
$2 \times (-0.61)^2 - 7(-0.61) - 5$
$= 0$ ✓

Finding Approximate Solutions to Equations

Some equations do not have exact solutions and cannot be solved by an algebraic method. Approximate solutions can be found by trial and improvement and iteration.

For **trial and improvement**, start by estimating a solution, then increasing it (if it is too small) or decreasing it (if it is too big) until you get as close as possible to the solution.

For **iteration**, repeat the same instructions until you find an accurate solution. This method is useful when solving cubic equations.

Example

Solve the equation $x^3 + 2x = 58$, by iteration, giving your answer to 5 decimal places. Start with the value $x = 3$.

$x^3 + 2x = 58$

$x^3 = 58 - 2x$

$x = \sqrt[3]{58 - 2x}$

First rearrange the equation to make x the subject. (You may be given this in a rearranged form already).

Now apply the iteration.

$x_{(n+1)} = \sqrt[3]{58 - 2x_n}$

$x_0 = 3$

$x_1 = \sqrt[3]{58 - 6} = 3.7325111\ldots$

$x_2 = \sqrt[3]{58 - 7.465022\ldots} = 3.697124\ldots$

$x_3 = \sqrt[3]{58 - 7.394248\ldots} = 3.698849\ldots$

$x_4 = \sqrt[3]{58 - 7.397698\ldots} = 3.698765\ldots$

$x_5 = \sqrt[3]{58 - 7.397530\ldots} = 3.698769\ldots$

$x_6 = \sqrt[3]{58 - 7.397538\ldots} = 3.698769\ldots$

$x = \mathbf{3.69877}$ — Round the value to 5 decimal places.

Substituting $x = 3.69877$ into $x^3 + 2x$ gives $3.69877^3 + (2 \times 3.69877) = 58.00004069$, which is very close to 58.

QUESTIONS

QUICK TEST

1. Solve:
   a. $x^2 - 6x + 8 = 0$ b. $x^2 + 5x + 4 = 0$
   c. $x^2 - 4x - 12 = 0$

2. Solve these equations by **i.** factorisation **ii.** using the quadratic formula.
   a. $x^2 + 2x - 15 = 0$ b. $2x^2 + 5x + 2 = 0$

EXAM PRACTICE

1. Charles cuts a square out of a rectangular piece of card.

   The length of the rectangle is $2x + 5$
   The width of the rectangle is $x + 3$
   The length of the side of the square is $x + 1$
   All measurements are in centimetres.

   The shaded shape in the diagram shows the card remaining.

   The area of the shaded shape is 45 cm^2.

   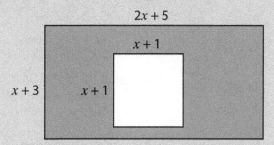

   a. Show that $x^2 + 9x - 31 = 0$ [4 marks]
   b. i. Solve the equation $x^2 + 9x - 31 = 0$
      Give your answer correct to 3 significant figures.   [3 marks]
      ii. Hence, find the perimeter of the square. Give your answer correct to 3 significant figures.   [1 mark]

2. An approximate solution to an equation is found by using this iterative process:

   $x_{n+1} = \sqrt[3]{40 - x_n}$

   $x_0 = 3$

   Work out the solution to 5 decimal places.   [3 marks]

Completing the Square

Completing the square is when **quadratic expressions**, $ax^2 + bx + c$, are written in the form $(x + p)^2 + q$.

To complete the square of the expression $ax^2 + bx + c$:

● If a is not 1, divide the whole expression by a.

● Write the expression in the form $\left(x + \dfrac{b}{2}\right)^2$

  Notice that the number in the bracket is always half the value of b, the coefficient of x.

● Multiply out the brackets, compare to the original and adjust by adding or subtracting an extra amount.

● Check: the value of $p = \dfrac{b}{2}$ and $q = -\left(\dfrac{b}{2}\right)^2 + c$

Example

The expression $x^2 + 8x + 7$ can be written in the form $(x + p)^2 + q$ for all values of x.

a. Find the values of p and q.

  $x^2 + 8x + 7$ is in the form $ax^2 + bx + c$

  $(x + 4)^2$ ← Half of 8 is 4

  $(x + 4)(x + 4)$
  $= x^2 + 8x + 16$ ← Multiply out the brackets and now compare to the original $x^2 + 8x + 7$

  $(x + 4)^2 - 9$ ← To make the expression equal, you subtract 9

  Hence $p = \mathbf{4}$ and $q = \mathbf{-9}$

  Check: $p = \dfrac{b}{2}$ $q = -\left(\dfrac{b}{2}\right)^2 + c$

  $p = \dfrac{8}{2}$ $q = -\left(\dfrac{8}{2}\right)^2 + 7$

  $p = 4$ $q = -(4)^2 + 7$

  $q = -16 + 7$

  $q = -9$

---

b. The expression $x^2 + 8x + 7$ has a minimum value. Find this minimum value.

  Since $(x + 4)^2 \geqslant 0$ for all values of x, the minimum value is equal to q.

  Minimum value $= \mathbf{-9}$, this occurs when $x = -4$.

Turning Points

Every quadratic graph has a turning point at a minimum if $a > 0$ or a maximum if $a < 0$.

For the quadratic equation $x^2 + 8x + 7$, the turning point, a minimum, has been found by completing the square and the coordinates of the turning point are $(-4, -9)$.

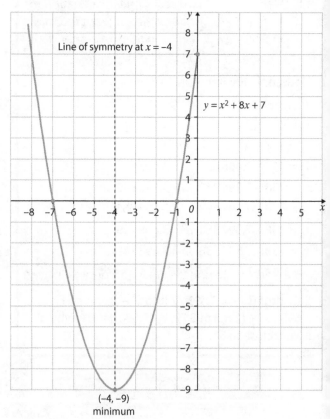

From the graph it can be seen that there is a line of symmetry, when $x = -4$. For a quadratic curve $y = ax^2 + bx + c$, the line of symmetry is $x = -\dfrac{b}{2a}$. Both halves of the curve are identical either side of the line.

Example

a. Write $2x^2 + 4x + 5$ in the form $a(x + b)^2 + c$, where a, b and c are integers.

$2x^2 + 4x + 5$

| Divide each term in the equation by 2. |

$2\left[x^2 + 2x + \dfrac{5}{2}\right]$

| Complete the square to get in the form $a(x + b)^2 + c$ |

$2\left[(x + 1)^2 + \dfrac{5}{2}\right]$

$2\left[(x^2 + 2x + 1) + \dfrac{5}{2} - 1\right]$

| Multiply out the brackets, compare to the original and adjust. |

$2\left[(x + 1)^2 + \dfrac{3}{2}\right]$

$\mathbf{2(x + 1)^2 + 3}$

| Multiply out by 2 to give this. |

b. Hence, write down the coordinates of the turning point of the graph $y = 2x^2 + 4x + 5$

$y = 2(x + 1)^2 + 3$

| Minimum occurs when $x = -1$ and $y = (2 \times 0) + 3$, so $y = 3$ |

Coordinates are **(−1, 3)**

Solving Quadratic Equations by Completing the Square

● Rearrange the equation into the form $ax^2 + bx + c = 0$
● If a is not 1, divide every term by a.
● Complete the square to get $(x + p)^2 + q = 0$
● Solve $(x + p)^2 = -q$

Example

Solve $x^2 = 6x - 2$ using the method of completing the square, giving your answers correct to 2 decimal places.

$x^2 = 6x - 2$

| Rearrange in the form $ax^2 + bx + c = 0$ |

$x^2 - 6x + 2 = 0$

| Half of −6 is −3 |

$(x - 3)^2$

| Multiply out the brackets and now compare to the original $x^2 - 6x + 2$ |

$(x - 3)^2 = x^2 - 6x + 9$

$(x - 3)^2 - 7 = 0$

| To make the expression equal, you subtract 7 |

$(x - 3)^2 - 7 = x^2 - 6x + 2$

$\therefore (x - 3)^2 - 7 = 0$

Now solve:

$(x - 3)^2 = 7$

| Square root both sides. Remember ±. |

$(x - 3) = \pm \sqrt{7}$

$x = 3 \pm \sqrt{7}$

| Add 3 to both sides. |

$\therefore x = 3 + \sqrt{7} = 5.645\ldots$ or $x = 3 - \sqrt{7} = 0.354\ldots$

$x = \mathbf{5.65}$ (2 d.p.) or $x = \mathbf{0.35}$ (2 d.p.)

SUMMARY

● **Completing the square is the method used when quadratic equations are expressed in the form $(x + p)^2 + q = 0$**

● **The coordinates of the turning point of $(x + p)^2 + q$ are at $(-p, q)$.**

QUESTIONS

QUICK TEST

1. The expression $x^2 - 4x + 7$ can be written in the form $(x + p)^2 + q$ for all values of x. Find the values of p and q.

2. The expression $x^2 + 6x + 5$ can be written in the form $(x + d)^2 + e$, for all values of x. Find the values of d and e.

EXAM PRACTICE

1. The expression $x^2 + 10x + 5$ can be written in the form $(x + a)^2 + b$ for all values of x.

   a. Find a and b. [3 marks]

   b. The expression $x^2 + 10x + 5$ has a minimum value. Find this minimum value. [1 mark]

2. **a.** Write $2x^2 + 8x + 3$ in the form $a(x + b)^2 + c$, where a, b and c are integers. [3 marks]

   b. Hence, write down the coordinates of the turning point of the graph $y = 2x^2 + 8x + 3$. [1 mark]

Simultaneous Linear Equations

Two equations with two unknowns are called **simultaneous equations**. They can be solved algebraically or graphically.

Solving Algebraically (Elimination Method)

Solve simultaneously:	$3x + 2y = 8$ $2x - 3y = 14$

Label the equations ① and ②.	$3x + 2y = 8$ ① $2x - 3y = 14$ ②

Since no coefficients match, multiply equation ① by 2 and equation ② by 3.	$6x + 4y = 16$ $6x - 9y = 42$

Rename them equations ③ and ④.	$6x + 4y = 16$ ③ $6x - 9y = 42$ ④

The coefficient of x in equations ③ and ④ is the same. Subtract equation ④ from equation ③ and solve to find y.	$0x + 13y = -26$ $y = -26 \div 13$ $y = -2$	Note $4y - (-9y)$ $= 4y + 9y$ $= 13y$

Substitute the value of $y = -2$ into equation ①. Solve this equation to find x.	$3x + 2 \times (-2) = 8$ $3x + (-4) = 8$ $3x = 8 + 4$ $3x = 12$ $x = 4$	You could substitute into ②.

Check in equation ②.	$(2 \times 4) - (3 \times -2) = 14$ ✓

Solution is: $x = 4$, $y = -2$

Solving Graphically

The point at which any two graphs **intersect** represents the simultaneous solution of their equations.

Example

Solve the simultaneous equations:

$2x + 3y = 6$

$x + y = 1$

Draw the graph of:

$2x + 3y = 6$

When $x = 0$, $3y = 6$ ∴ $y = 2$ (0, 2)

When $y = 0$, $2x = 6$ ∴ $x = 3$ (3, 0)

Draw the graph of:

$x + y = 1$

When $x = 0$, $y = 1$ (0, 1)

When $y = 0$, $x = 1$ (1, 0)

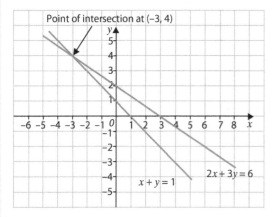

At the point of intersection: $x = -3$, $y = 4$. This is the solution of the simultaneous equations.

SUMMARY

- When you are solving simultaneous equations, there are two equations with two unknowns.

- The elimination method involves eliminating one of the variables.

- The point where two graphs cross gives the solution to both equations.

QUESTIONS

QUICK TEST

1. Solve the simultaneous equations:

   $4b + 7a = 10$

   $2b + 3a = 3$

2. The diagram shows the graphs of the lines:

   $x + y = 6$ and $y = x + 2$

   Use the diagram to solve the simultaneous equations $x + y = 6$ and $y = x + 2$.

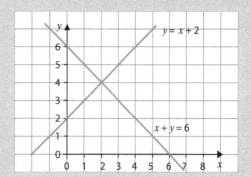

EXAM PRACTICE

1. Solve the simultaneous equations:

   $5a - 2b = 19$

   $3a + 4b = 1$ [4 marks]

2. Frances and Patrick were organising a children's party. They went to the toy shop and bought some hats and balloons. Frances bought five hats and four balloons. She paid £22. Patrick bought three hats and five balloons. He paid £21.

   The cost of a hat was x pounds. The cost of a balloon was y pounds.

   Work out the cost of one hat and the cost of one balloon.   [5 marks]

Algebraic Fractions

Simplifying Algebraic Fractions

When working with **algebraic fractions**, the same rules are used as with ordinary fractions.

Example

Simplify $\dfrac{8x + 16}{x^2 - 4}$

First, you need to factorise the numerator and denominator.

$\dfrac{8(x + 2)}{(x + 2)(x - 2)}$

Remember the difference of two squares:
$a^2 - b^2 = (a + b)(a - b)$

Now cancel any common factors, i.e. the $(x + 2)$.

$\dfrac{8\cancel{(x + 2)}}{\cancel{(x + 2)}(x - 2)} = \dfrac{8}{x - 2}$

Multiplication

When multiplying algebraic fractions, multiply the **numerators** and multiply the **denominators** and then cancel if possible.

Example

Solve $\dfrac{3a^2}{4b} \times \dfrac{16b^2}{9ab}$

$= \dfrac{48a^2b^2}{36ab^2}$

$= \dfrac{4a}{3}$

Division

When dividing algebraic fractions, remember to turn the second fraction upside down (i.e. take the **reciprocal**) and then multiply and cancel if possible.

Example

Solve $\dfrac{12(x - 3)}{(x + 2)} \div \dfrac{4(x - 1)(x - 3)}{(x + 1)}$

Cancel out common factors.

$= \dfrac{^3\cancel{12}\,\cancel{(x - 3)}}{(x + 2)} \times \dfrac{(x + 1)}{\cancel{4}(x - 1)\cancel{(x - 3)}}$

$= \dfrac{3(x + 1)}{(x + 2)(x - 1)}$

Addition and Subtraction

As with ordinary fractions, you cannot add or subtract unless the fractions have the same **denominator**.

Example

Solve $\dfrac{(x + 3)}{(x + 2)} + \dfrac{3}{(x - 1)}$

The common denominator is:
$(x + 2)(x - 1)$

The numerator now needs to be adjusted.

$\dfrac{(x + 3)(x - 1) + 3(x + 2)}{(x + 2)(x - 1)}$

Multiply out the numerator.

$\dfrac{x^2 + 2x - 3 + 3x + 6}{(x + 2)(x - 1)}$

Simplify, then check whether the numerator can by factorised and the expression simplified further by cancelling.

$= \dfrac{x^2 + 5x + 3}{(x + 2)(x - 1)}$

Solving Equations with Algebraic Fractions

Equations with algebraic fractions sometimes lead to quadratic equations, which can then be solved as shown previously.

Example

Solve $\dfrac{3}{x+1} + \dfrac{4}{x+2} = 2$

> The common denominator is $(x+1)(x+2)$. Now adjust the numerator.

$$\frac{3(x+2) + 4(x+1)}{(x+1)(x+2)} = 2$$

> Expand the brackets and simplify.

$$\frac{3x+6+4x+4}{x^2+3x+2} = 2$$

> Multiply the right-hand side by $x^2 + 3x + 2$

$$7x + 10 = 2(x^2 + 3x + 2)$$

$$7x + 10 = 2x^2 + 6x + 4$$

> Make the equation equal to zero.

$$2x^2 - x - 6 = 0$$

> Factorise, then solve the quadratic equation.

$$(2x + 3)(x - 2) = 0$$

So either $x = -\dfrac{3}{2}$ or $x = 2$

SUMMARY

- **With algebraic fractions, the same rules are applied as with ordinary fractions.**
  - **When adding or subtracting, make sure that the denominators are the same.**
  - **When multiplying, multiply the numerators and multiply the denominators.**
  - **When dividing, turn the second fraction upside down (take the reciprocal) and then multiply.**

QUESTIONS

QUICK TEST

1. Decide whether these expressions are fully simplified:

   a. $\dfrac{6s^2w}{3w^2}$ b. $\dfrac{5a^2b}{2c}$ c. $\dfrac{9abc}{d}$

2. Simplify the following algebraic fractions:

   a. $\dfrac{3}{(x+1)} + \dfrac{2}{(x-1)}$

   b. $\dfrac{3a^2b}{2q} \times \dfrac{4q^2}{9abc}$

   c. $\dfrac{10(a^2 - b^2)}{3x + 6} \div \dfrac{(a - b)}{4x + 8}$

EXAM PRACTICE

1. Simplify fully: $\dfrac{8x + 16}{x^2 - 4}$ [2 marks]

2. Simplify fully: $\dfrac{x^2(6 + x)}{x^2 - 36}$ [2 marks]

3. Simplify fully: $\dfrac{2x^2 + 7x - 15}{x^2 + 3x - 10}$ [3 marks]

4. Solve: $\dfrac{3}{x} - \dfrac{4}{x + 1} = 1$ [5 marks]

Sequences

A **sequence** is a set of numbers that follow a particular rule. The word '**term**' is often used to describe each number in the sequence.

A term-to-term sequence means you can find a rule for each term based on the previous term in the sequence. For example, in the sequence 2, 4, 6, 8 … you add 2 each time to go from one term to the next.

Special Sequences

Odd numbers   1, 3, 5, 7, 9 … nth term is $2n - 1$
Even numbers   2, 4, 6, 8, 10 … nth term is $2n$

Square Numbers

  1   4   9

$1^2 = 1 \times 1$ $2^2 = 2 \times 2$ $3^2 = 3 \times 3$

  16   25   …

> You need to know all the square numbers up to 15^2.

$4^2 = 4 \times 4$ $5^2 = 5 \times 5$

Cube Numbers

  1   8   27   64   125…

$1^3 = 1 \times 1 \times 1$ $2^3 = 2 \times 2 \times 2$ $3^3 = 3 \times 3 \times 3$

Triangular Numbers

  1   3   6   10   15   …

    +2   +3   +4   +5

Fibonacci Sequence

1, 1, 2, 3, 5, 8, 13 …   Add the previous two terms.

There are other types of Fibonacci sequence. One is the Lucas series: 2, 1, 3, 4, 7, 11, 18 …

Finding the nth Term of an Arithmetic Sequence

The nth term is the rule for a sequence and is often denoted by U_n. For example, the 8th term is U_8.

For a linear or arithmetic sequence, the nth term takes the form:

$$U_n = an + b$$

Example
Find the nth term of this sequence: 2, 6, 10, 14 …

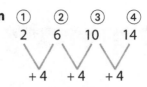

Position ①   ②   ③   ④
Term   2   6   10   14

         + 4   + 4   + 4

- See how the numbers are jumping (going up in 4s).

- The nth term is $4n +$ or $-$ something.

- Try $4n$ on the first term. This gives $4 \times 1 = 4$, but the first term is 2, so subtract 2.

- The rule is $4n - 2$

- Test this rule on the other terms:

  $1 \rightarrow (4 \times 1) - 2 = 2$

  $2 \rightarrow (4 \times 2) - 2 = 6$

  $3 \rightarrow (4 \times 3) - 2 = 10$

  It works on all of them.

- nth term is **$4n - 2$**

  The 20th term in the sequence would be:

  $(4 \times 20) - 2 = 78$

Geometric Sequences

In a geometric sequence, you multiply by a constant number (common ratio), r, to go from one term to the next.

For example, consider this sequence:

    5   10   20   40   80

The term-to-term formula is $U_{n+1} = 2U_n$, where $U_1 = 5$

The position-to-term formula is $U_n = 5 \times 2^{n-1}$

Finding the nth Term of a Quadratic Sequence

For a quadratic sequence, the first differences are not constant but the second differences are. The nth term U_n takes the form $U_n = an^2 + bn + c$, where b and c may be zero.

Examples

1. Find an expression for the nth term for the sequence of square numbers.

Term	1	2	3	4	5
Sequence	1	4	9	16	25

First difference   3   5   7   9

Second difference   2   2   2

Since the second differences are the same, the rule for the nth term is quadratic. In this case, the nth term is $\mathbf{n^2}$.

2. Find the nth term of the sequence 3, 9, 19, 33, 51 ...

Term	1	2	3	4	5
Sequence	3	9	19	33	51

First difference   6   10   14   18

Second difference   4   4   4

Since the second differences are the same, the rule for the nth term is quadratic. The coefficient of n^2 is (the second difference) $\div 2$, i.e. $4 \div 2 = 2$ Adjusting as with linear sequences gives $\mathbf{2n^2 + 1}$.

SUMMARY

● A sequence is a set of numbers that follow a particular rule.

● The nth term is the rule for a sequence and is denoted by U_n.

● Given a sequence, you will need to be able to work out the nth term.

QUESTIONS

QUICK TEST

1. The cards show the nth term of some sequences:

$2n$	$4n+1$	$3n+2$	$5n-1$	$2-n$

Match the cards with the sequences below:

a. 5, 9, 13, 17 ...   **b.** 1, 0, −1, −2 ...

c. 2, 4, 6, 8, 10 ...   **d.** 5, 8, 11, 14, 17 ...

e. 4, 9, 14, 19 ...

EXAM PRACTICE

1. Here are the first four terms of an arithmetic sequence:

    5   7   9   11

Find an expression in terms of n for the nth term of the sequence.   [2 marks]

2. 🚫 The nth term of a sequence is $2n^2 + 1$. Chloe says that 101 is a number in the sequence.

Explain whether Chloe is correct.   [2 marks]

3. Write down the formula for the nth term of this sequence.

    7, 22, 47, 82, 127 ...   [3 marks]

Inequalities

An **inequality** is a statement showing two quantities that are not equal. Inequalities can be solved in the same way as equations, except that when multiplying or dividing by a negative number you must reverse the inequality sign.

The Inequality Symbols

$>$ means **greater than**
$<$ means **less than**
$\geq$ means **greater than or equal to**
$\leq$ means **less than or equal to**

Examples
Solve:

1. $2x - 2 < 10$

$2x < 10 + 2$

$2x < 12$

$\boldsymbol{x < 6}$

2. $3 - 2x \geq 9$

$-2x \geq 9 - 3$

$-2x \geq 6$

$x \leq \dfrac{6}{-2}$ ← | Divide by −2 and reverse the inequality. |

$\boldsymbol{x \leq -3}$

| Add 1 to each part of the inequality. |

3. $-7 < 3x - 1 \leq 11$

$-6 < 3x \leq 12$ ←

$\boldsymbol{-2 < x \leq 4}$ ← | Divide each part of the inequality by 3. |

The integer values that satisfy this inequality are −1, 0, 1, 2, 3, 4.

Number Lines

Inequalities can be shown on a number line.

$x < 6$

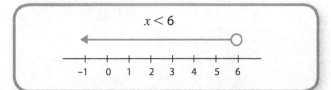

The open circle means that 6 is not included.

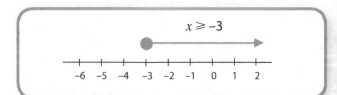

The solid circle means that −3 is included.

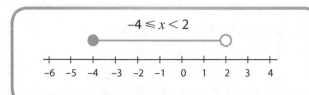

The integer values that satisfy this inequality are −4, −3, −2, −1, 0, 1.

Graphs of Inequalities

The graph of an equation such as $x = 2$ is a line, whereas the graph of the inequality $x < 2$ is a **region** that has $x = 2$ as its boundary.

The diagram below shows unshaded the region R:

$x + y \leq 7$ $x > 2$ $y \geq 2$

For strict inequalities $>$ and $<$ the boundary line is not included and is shown as a dashed line.

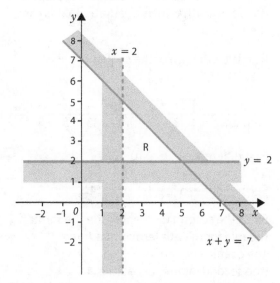

Quadratic Inequalities

To solve a quadratic inequality you:
- Write the quadratic inequality in the form of the quadratic equation: $ax^2 + bx + c = 0$
- Solve the corresponding quadratic equation
- Sketch the graph of the quadratic function
- Use the sketch to find the required set of values.

Examples

1. Find the set of values of x for which $x^2 - 3x - 4 < 0$ and draw a sketch to show this.
   - Write the quadratic inequality in the form of the quadratic equation: $ax^2 + bx + c = 0$
     $x^2 - 3x - 4 = 0$
   - Factorise the quadratic equation:
     $(x - 4)(x + 1) = 0$
   - Solve the quadratic equation:
     Either $x - 4 = 0$, so $x = 4$
     Or $x + 1 = 0$, so $x = -1$
   - Sketch the quadratic graph. The sketch does not need to be accurate. All you need to know is that the graph is ∪ shaped and crosses the x-axis at $x = -1$ and $x = 4$

   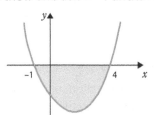

   $x^2 - 3x - 4 < 0$ ($y < 0$) for the part of the graph below the x-axis, as shown by the part shaded blue on the diagram.

   The required set of values is **$-1 < x < 4$**.

2. Find the set of values of x for which $x^2 - 3x - 4 > 0$

   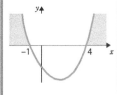

   The only difference between this example and the previous one is the final stage, since you want the set of values for which $x^2 - 3x - 4 > 0$

   $x^2 - 3x - 4 > 0$ ($y > 0$) for the part of the graph above the x-axis.

   The required set of values is **$x < -1$ or $x > 4$**

SUMMARY

- $>$ greater than
  $<$ less than
  $\geq$ greater than or equal to
  $\leq$ less than or equal to

- On number lines, an open circle means the value is not included in the inequality and a solid circle means the value is included in the inequality.

- On graphs of inequalities, use a dashed line when the boundary is not included.

QUESTIONS

QUICK TEST

1. Solve the following inequalities:
   a. $5x - 1 < 10$ b. $6 \leq 3x + 2 < 11$
   c. $3 - 5x < 12$

EXAM PRACTICE

1. n is an integer such that $-6 < 2n \leq 8$.
   List all the possible values of n. [1 mark]

2. Solve the inequality $4 + x > 7x - 8$. [2 marks]

3. On the diagram below, leave unshaded the region satisfied by these inequalities:
   $x + y \leq 5$
   $x \geq 1$
   $y > 1$
   [3 marks]

   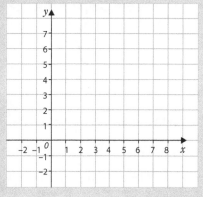

4. Solve $x^2 > 5x + 6$ [3 marks]

Straight-line Graphs

Drawing Straight-line Graphs

The general equation of a straight line is:

$$y = mx + c$$

m is the **gradient**. c is the **intercept** on the y-axis.

To draw the graph of $y = 3x - 4$:

⬤ Work out the coordinates of the points that lie on the line $y = 3x - 4$ by drawing a table of values for x. Substitute the x values into the equation $y = 3x - 4$, to find the values of y.

e.g. $x = 2$, $y = 3 \times 2 - 4 = 2$

x	−1	0	2	4
y	−7	−4	2	8

⬤ The coordinates of the points on the line are:

(−1, −7)   (0, −4)   (2, 2)   (4, 8)

Just read them from the table of values.

⬤ Plot the points (across the whole grid) and join with a straight line.

⬤ The line $y = 3x - 4$ is drawn.

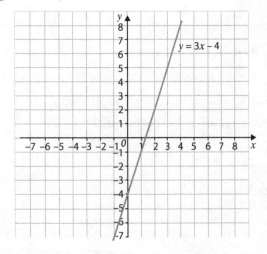

⬤ Label the line once you have drawn it.

Gradient of a Straight Line

Be careful when finding the gradient: double-check the scales.

$$\text{Gradient} = \frac{\text{change in } y}{\text{change in } x}$$

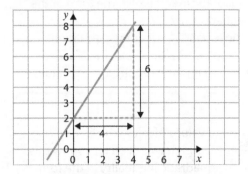

For the straight line above:

$$\text{Gradient} = \frac{6}{4} = \frac{3}{2} = 1.5$$

Equation of the line is $y = 1.5x + 2$

Positive and Negative Gradients

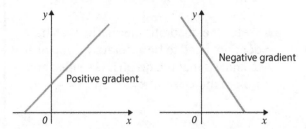

Positive gradient

Negative gradient

Finding the Midpoint of a Line Segment

The midpoint of a line segment between two points can be worked out by finding the mean of the x coordinates and the mean of the y coordinates of the points.

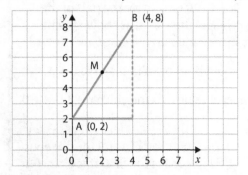

The midpoint of a line that joins the point A(x_1, y_1) and B(x_2, y_2) is:

$$\left(\frac{(x_1 + x_2)}{2}, \frac{(y_1 + y_2)}{2} \right)$$

The midpoint, M, of the line AB drawn here is:

$$\left(\frac{(0 + 4)}{2}, \frac{(2 + 8)}{2} \right) = (2, 5)$$

Perpendicular Lines

If two lines are **perpendicular** the product of their gradients is –1.

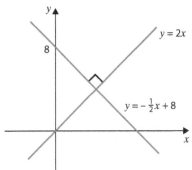

The lines $y = 2x$ and $y = -\frac{1}{2}x + 8$ are perpendicular because the gradients multiply to give –1.

$$\left(2 \times -\frac{1}{2} = -1\right)$$

Finding the Equation of Straight Lines

You need to be able to find the equations of both parallel and perpendicular lines when key information is given. Remember the equation of a straight line is in the form $y = mx + c$.

Examples

1. Find the equation of a line with gradient –3 going through the point (–2, 10).

The equation of the line is $y = mx + c$

$10 = (-3 \times -2) + c$ ⟵ | Substitute the gradient and coordinates into the equation; use $x = -2$ and $y = 10$

$c = 4$

Equation of the line is $y = -3x + 4$

2. A line L is perpendicular to the line $y = 4x - 2$ and passes through the point (12, 5). Find the equation of L.

The gradient of $y = 4x - 2$ is 4. So the gradient of a line perpendicular to it has a gradient of $-\frac{1}{4}$

Substitute $x = 12$, $y = 5$ and $m = -\frac{1}{4}$ into $y = mx + c$

$5 = -\frac{1}{4} \times 12 + c$

$c = 8$

Equation of L is $y = -\frac{1}{4}x + 8$

QUESTIONS

QUICK TEST

1. a. Complete the table of values for $y = 2x + 3$.

x	–2	–1	0	1	2	3
y						

b. Draw the graph of $y = 2x + 3$.

EXAM PRACTICE

1.

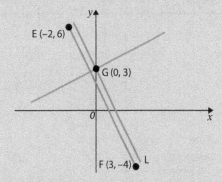

The diagram shows three points: E (–2, 6), F (3, –4) and G (0, 3). A line L is parallel to EF and passes through G.

a. Find an equation for the line L. [3 marks]

b. Find the midpoint of the line EF. [2 marks]

c. Find the equation of a line perpendicular to line L and which passes through point G. [2 marks]

Curved Graphs

Quadratic Graphs

Quadratic graphs are of the form $y = ax^2 + bx + c$ where $a \neq 0$. Quadratic graphs have an x^2 term as the highest power of x. They will be ∪ shaped if the **coefficient** of x^2 is positive, and ∩ shaped if the coefficient of x^2 is negative.

To draw the graph of $y = x^2 - 2x - 6$ using values of x from −2 to 4:

- Draw a table of values. Fill in the table of values by substituting the values of x into the equation.

  e.g. $x = 1,$ $y = 1^2 - 2 \times 1 - 6 = -7$
  Coordinates are $(1, -7)$

x	−2	−1	0	1	2	3	4
y	2	−3	−6	−7	−6	−3	2

- Draw the axes on graph paper and plot the points.
- Join the points with a smooth curve.
- Label the curve.

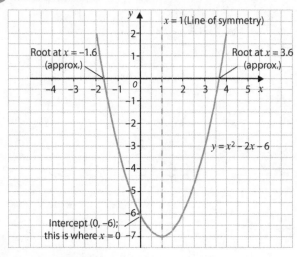

The **minimum** point is $(1, -7)$.
The **line of symmetry** is $x = 1$.

The **intercept** on the y-axis is the point at which $x = 0$. On the graph $y = x^2 - 2x - 6$, the intercept on the y-axis is $(0, -6)$.

The **roots** of the quadratic equation occur when $y = 0$, i.e. where the graph crosses the x-axis.

The roots of the graph $y = x^2 - 2x - 6$ are approximately $(-1.6, 0)$ and $(3.6, 0)$.

Tangents and Gradients

For a curved graph, the **gradient** is constantly changing. In order to find the gradient of a particular point on a curved graph, a **tangent** needs to be drawn at that point. The gradient of the tangent drawn is then worked out.

> **Example**
>
> - Draw a tangent so that it just touches the curve at that point.
> - Choose two points on the tangent and draw a triangle. Do not count the squares as the scales may be different.
> - Gradient $= \dfrac{\text{change in } y}{\text{change in } x}$
> - Decide if the gradient is positive or negative.
>
> $$\text{Gradient} = \frac{\text{change in } y}{\text{change in } x} = \frac{19 - 1}{4.8 - 1.6} = \frac{18}{3.2} = 5.625$$
>
>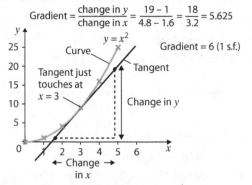

Trigonometric Functions

The behaviour of the sine, cosine and tangent functions may be represented graphically.

> **$y = \sin x$**   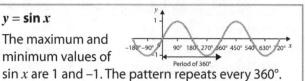
>
> The maximum and minimum values of $\sin x$ are 1 and −1. The pattern repeats every 360°.

> **$y = \cos x$**   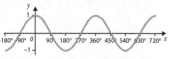
>
> The maximum and minimum values of $\cos x$ are 1 and −1. The pattern repeats every 360°. This graph is the same as $y = \sin x$ except it has been translated 90° to the left.

$y = \tan x$

This graph is nothing like the sine or cosine curve. The values of tan x repeat every 180°.

The tan of 90° is infinity, i.e. a value so great it cannot be written down.

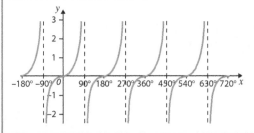

Graph Shapes

$y = mx + c$

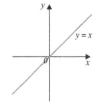

$y = ax^2 + bx + c$

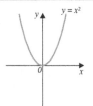

$y = \dfrac{k}{x}$

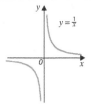

$y = x^3$

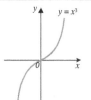

$y = a^x$

SUMMARY

- Make sure you know the different graph shapes.

- Quadratic graphs have an x^2 term as the highest power of x. They are $\cup$ shaped or $\cap$ shaped.

- Cubic graphs have an x^3 term as the highest power of x.

- The reciprocal function is $y = \dfrac{k}{x}$

QUESTIONS

QUICK TEST

1. Match each graph below to one of the equations.

$$y = x^3 - 5 \qquad y = 2 - x^2 \qquad y = 4x + 2 \qquad y = \dfrac{3}{x}$$

Graph A

Graph B

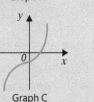

Graph C

Graph D

EXAM PRACTICE

1. **a.** Complete the table of values for $y = x^3 - 1$

x	−3	−2	−1	0	1	2	3
y			−1				

[2 marks]

b. Using a suitable scale, draw the graph of $y = x^3 - 1$. [3 marks]

c. From the graph find the approximate value of x when $y = 15$. [2 marks]

d. From your graph find the gradient at $x = 2$. [3 marks]

Solving a Linear and a Non-linear Equation Simultaneously

You need to be able to work out the **coordinates** of the points of intersection of a straight line and a quadratic curve, as well as a straight line and a circle.

Point of Intersection of a Straight Line and a Quadratic Graph

Example

a. Solve the simultaneous equations:

$y = 4x - 2$ ①

$y = x^2 + 1$ ②

> Eliminate y by substituting equation ② into equation ①.

$x^2 + 1 = 4x - 2$

Rearrange: $x^2 - 4x + 3 = 0$

Factorise: $(x - 1)(x - 3) = 0$

Solve: $x = 1$ and $x = 3$

> Now substitute the values of x back into equation ② to find the corresponding values of y.

When $x = 1$, $y = 1^2 + 1 = 2$;
$x = \mathbf{1}$, $y = \mathbf{2}$

When $x = 3$, $y = 3^2 + 1 = 10$;
$x = \mathbf{3}$, $y = \mathbf{10}$

b. Give a geometrical interpretation of the result.

The diagram shows the points of intersection, A and B, of the line $y = 4x - 2$ and the curve $y = x^2 + 1$. You know that the points A and B lie on both the curve and the line.

Coordinates are A (1, 2) and B (3, 10).

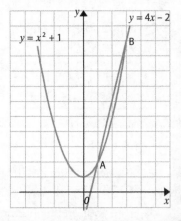

Points of Intersection of a Straight Line and a Circle

The equation of a circle with centre (0, 0) and radius r is $x^2 + y^2 = r^2$

Example

Find the coordinates of the points where the line $y - x = 6$ intersects $x^2 + y^2 = 36$.

The coordinates must satisfy both equations, so you solve them simultaneously.

$y - x = 6$ ①

$x^2 + y^2 = 36$ ②

> Rewrite equation ① in the form '$y =$'

$y = x + 6$ ①

> Eliminate y by substituting equation ① into equation ②.

$x^2 + (x + 6)^2 = 36$

> Expand brackets: $(x + 6)^2 = (x + 6)(x + 6)$
> $= x^2 + 12x + 36$

$x^2 + x^2 + 12x + 36 = 36$

Rearrange: $2x^2 + 12x = 0$

Factorise: $2x(x + 6) = 0$

Solve: $x = 0$, $x = -6$

> Substitute the values of x into equation ① to find the values of y.

Since $y - x = 6$, $y = x + 6$

When $x = 0$, $y = 0 + 6 = 6$

When $x = -6$, $y = -6 + 6 = 0$

The line $y - x = 6$ intersects the circle $x^2 + y^2 = 36$ at (0, 6) and (−6, 0).

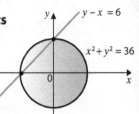

Circles and Tangents

A **tangent** to a circle is a straight line that touches the circumference at one point only. The tangent to a circle and the radius meet at 90°, hence the radius and tangent line are **perpendicular** to each other. If the radius has gradient m, then the tangent line has gradient $-\frac{1}{m}$

Example

Work out the equation of the tangent to the circle $x^2 + y^2 = 29$ at the point (2, 5).

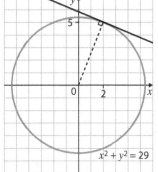

The gradient (m) of the radius is $\frac{5}{2}$

The gradient of the perpendicular line, i.e. the tangent, is $-\frac{1}{m}$, hence $-\frac{2}{5}$

$5 = \left(-\frac{2}{5} \times 2\right) + c$

$c = \frac{29}{5}$

Substitute $x = 2$, $y = 5$ and the gradient of $-\frac{2}{5}$ into $y = mx + c$ to work out c.

The equation of the tangent is $y = -\frac{2}{5}x + \frac{29}{5}$ or

$5y = 29 - 2x$

SUMMARY

- Two equations with two unknowns are called simultaneous equations.

- Simultaneous equations can be solved by a graphical method or by a method of elimination or substitution.

- The equation $x^2 + y^2 = r^2$ is an equation of a circle with centre (0, 0) and radius r.

- The tangent to a circle and the radius meet at 90°, hence the radius and tangent line are perpendicular to each other. If the radius has gradient m, then the tangent line has gradient $-\frac{1}{m}$

QUESTIONS

QUICK TEST

1. Solve these simultaneous equations:

   a. $y = x + 4$
   $y = x^2 + 2$

   b. $y = x + 1$
   $x^2 + y^2 = 25$

2. For each of the two questions above, give a geometrical interpretation of the result.

EXAM PRACTICE

1. Ⓝ Find the coordinates of the points where the line $y - 5 = x$ intersects the circle $x^2 + y^2 = 25$ [6 marks]

2. Ⓝ Solve the simultaneous equations:
   $x^2 + y^2 = 26$
   $y = 2x + 3$ [6 marks]

3. Work out the equation of the tangent to the circle $x^2 + y^2 = 25$ at the point (3, 4). [3 marks]

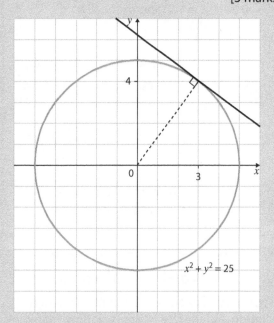

More Graphs and Other Functions

Using Graphs to Solve Equations

Often an equation needs to be rearranged in order to resemble the equation of the plotted graph.

Example

The graph of $y = x^2 - 4x + 4$ is drawn here:

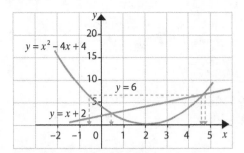

Use the graph to solve:

a. $x^2 - 4x + 4 = 6$

> The solution will be where the curve $y = x^2 - 4x + 4$ meets the line $y = 6$

$x = -0.5$ $x = 4.5$ (approximately)

b. $x^2 - 5x + 2 = 0$

> The equation needs to be rearranged so that one side is the same as the plotted graph $y = x^2 - 4x + 4$

$x^2 - 5x + 2 + x + 2 = x + 2$ ← Add $x + 2$ to both sides

$x^2 - 4x + 4 = x + 2$

> The solutions are where the curve $y = x^2 - 4x + 4$ crosses the line $y = x + 2$

$x = 0.4$ $x = 4.6$

Using Graphs to Find Relationships

Exponential graphs can help to estimate the rate of **appreciation** and the rate of **depreciation**. Examples are the exponential growth of bacteria and the exponential decay of radioactive elements.

Example

This graph is known to fit $y = pq^x$, where p and q are positive constants.

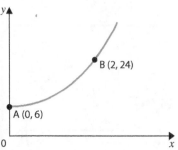

Use the graph to find the values of p and q and the relationship.

$6 = p \times q^0$ ← Substitute the coordinates of point A into $y = pq^x$

Since $q^0 = 1$ $p = 6$

$24 = 6 \times q^2$ ← Substitute the coordinates of point B into $y = pq^x$

$\therefore$ $4 = q^2$

$q = 2$

The relationship is $y = 6 \times 2^x$

Area Under a Curve

You can find the approximate area under a curved graph by splitting the area into trapeziums of equal width. The heights of the trapeziums are taken from the graph and the area of each trapezium is calculated.

Example

Find the area under the graph $y = x^2$ for values of x between 0 and 5.

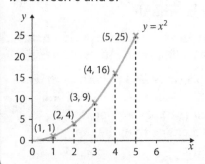

Example (cont.)

Split the graph into trapeziums of height 1. Then find the area of each trapezium:

Area of trapezium 1 = $\dfrac{(0+1)}{2}$

Area of trapezium 2 = $\dfrac{(1+4)}{2}$

Area of trapezium 3 = $\dfrac{(4+9)}{2}$

Area of trapezium 4 = $\dfrac{(9+16)}{2}$

Area of trapezium 5 = $\dfrac{(16+25)}{2}$

Total area = $\dfrac{1}{2}(1+5+13+25+41)$

= **approximately 42.5 squared units**

Speed–Time Graphs

A **speed–time** graph is useful when finding the **acceleration** or **deceleration** of an object. It can also be used to find the distance travelled.

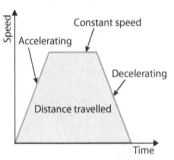

- Distance is the area between the graph and x-axis.

- A positive **gradient** means the speed is increasing.

- A negative **gradient** means the speed is decreasing.

- A horizontal line means the speed is constant.

SUMMARY

- Accurate graphs can be used to find solutions to some equations.

- Speed–time graphs can be used to find the distance travelled by an object and also acceleration and deceleration.

QUICK TEST

Draw the graph of $y = x^2 + x - 2$.
Use your graph to decide whether the solutions of these equations are true or false.

1. $x^2 + x = 5$
   solution is approximately $x = -2.8$, $x = 1.8$

2. $x^2 - 2 = 0$
   solution is approximately $x = 2.4$, $x = -2.4$

3. $x^2 - 2x - 2 = 0$
   solution is approximately $x = -0.7$, $x = 5.6$

EXAM PRACTICE

1. ⊘ Miss Jones has a car.

   The value of the car on 1 September 2014 was £9000. The value of the car on 1 September 2016 was £4000.

   The graph shows how the value, £v, of the car changes with time. The equation of the curve is:

   $v = ab^t$

   where t is the number of years after 1 September 2014, and a and b are positive constants.

   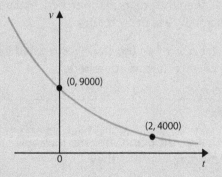

   Use the information on the graph to find the values of a and b. [3 marks]

Functions and Transformations

If y is equal to an expression involving x, then it can be written as $y = f(x)$. The graphs of the related **functions** can be found by applying **transformations**.

$y = f(x) \pm a$

This is when the graphs move up or down the y-axis by a value of a, i.e. a translation of $\begin{pmatrix} 0 \\ \pm a \end{pmatrix}$

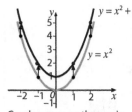

Graph moves up the y-axis by one unit.

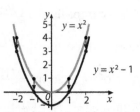

Graph moves down the y-axis by one unit.

$y = f(x \pm a)$

This is when the graphs move along the x-axis by a units.

$y = f(x + a)$ moves the graph a units to the left, i.e. a translation of $\begin{pmatrix} -a \\ 0 \end{pmatrix}$

$y = f(x - a)$ moves the graph a units to the right, i.e. a translation of $\begin{pmatrix} a \\ 0 \end{pmatrix}$

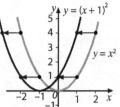

Graph of $y = (x + 1)^2$ moves one unit to the left.

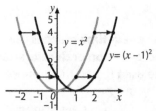

Graph of $y = (x - 1)^2$ moves one unit to the right.

$y = kf(x)$

This is when the original graph stretches along the y-axis by a factor of k.

If $k > 1$, then the points are stretched upwards in the y direction by a scale factor of k.

If $k < 1$, e.g. $y = \frac{1}{2}x^2$, the graph is stretched by a scale factor of $\frac{1}{2}$ (i.e. it looks squashed).

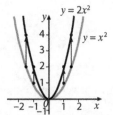

For this graph the x values stay the same and the y values are multiplied by 2.

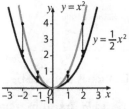

For this graph the x values stay the same and the y values are multiplied by $\frac{1}{2}$.

$y = f(kx)$

If $k > 1$, then the graph stretches inwards in the x direction by a scale factor $\frac{1}{k}$

If $k < 1$, then the graph stretches outwards in the x direction by a scale factor $\frac{1}{k}$

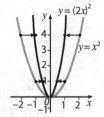

For this graph the y coordinates stay the same and the x values are multiplied by $\frac{1}{2}$.

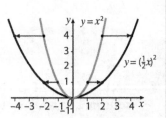

For this graph the y coordinates stay the same and the x values are multiplied by 2.

Applying Reflections

$y = -f(x)$
This is a reflection in the x-axis.

$y = f(-x)$
This is a reflection in the y-axis.

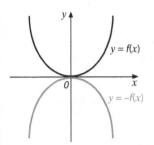

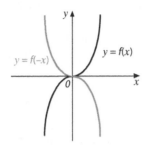

Composite and Inverse Functions

Composite functions consist of one or more functions.

For example, $f(x)$ and $g(x)$ are two functions where $f(x) = x - 6$ and $g(x) = x^2$

The composite function $fg(x)$ means do g first, then f. So the composite function is $fg(x) = x^2 - 6$

$fg(2) = 2^2 - 6$
$fg(2) = -2$ ← Simply substitute $x = 2$ into the equation.

A function maps one number to another number.

An **inverse** function maps one number back to the other. For example, if $f(x) = x - 6$, then the inverse function $f^{-1}(x) = x + 6$

SUMMARY

- If y is equal to an expression involving x, then it can be written as $y = f(x)$.

- The graphs of related functions can be found by applying transformations.

- Composite functions consist of one or more functions. If $f(x)$ and $g(x)$ are two functions, then the composite function $fg(x)$ means do g first, then f.

- A function maps one number to another number.

- An inverse function maps one number back to the other.

QUESTIONS

QUICK TEST

1. The graph of $y = f(x)$ is shown on the grid.

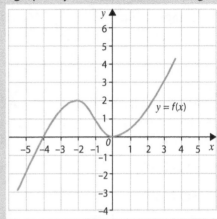

a. Sketch the graph of $y = f(x) + 1$

b. Sketch the graph of $y = -f(x)$

EXAM PRACTICE

1. 🚫 This is a sketch of the curve $y = f(x)$

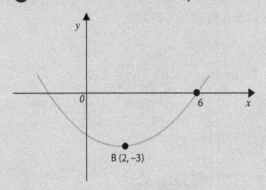

The minimum point of the curve is point B (2, −3).

Write down the coordinates of the minimum point for each of the following curves:

a. $y = f(x + 3)$ **b.** $y = f(x) - 4$

c. $y = f(x - 2)$ **d.** $y = f(-x)$

e. $y = f(x + 1)$ [5 marks]

2. If $f(x) = 2x + 3$ and $g(x) = x^2$ then work out:

a. $fg(x)$ [2 marks]   **b.** $f^{-1}(x)$ [2 marks]

Repeated Percentage Change

Percentage Change

> Percentage change = $\frac{change}{original} \times 100\%$

Examples

1. Tammy bought a flat for £185 000. Three years later she sold it for £242 000. What is her percentage profit?

Profit is £242 000 – £185 000

$= £57 000$

Percentage profit is $\frac{57\,000}{185\,000} \times 100\%$

$= \mathbf{30.8\%}$ (3 s.f.)

2. Jackie bought a car for £12 500 and sold it two years later for £7250. Work out her percentage loss.

Loss is £12 500 – £7250

$= £5250$

Percentage loss is $\frac{5250}{12\,500} \times 100\%$

$= \mathbf{42\%}$

Repeated Percentage Change

A quantity can increase or decrease in value each year by the same or a different percentage. These quantities will change in value at the end of each year. To calculate repeated percentage change, two methods are explained in the example below.

Example
A car was bought for £12 500. Each year it depreciated in value by 15%. What was the car worth after three years?

> You must remember **not** to do $3 \times 15\% = 45\%$ reduction over three years!

Method 1

● Find 100% – 15% = 85% of the value of the car first.

Year 1: $\frac{85}{100} \times £12\,500 = £10\,625$

● Then work out the value year by year. (£10 625 depreciates in value by 15%.)

Year 2: $\frac{85}{100} \times £10\,625 = £9031.25$

(£9031.25 depreciates in value by 15%.)

Year 3: $\frac{85}{100} \times £9031.25 = \mathbf{£7676.56}$

Method 2

● A quick way to work this out is by using a **multiplier**.

● Finding 85% of the value of the car is the same as multiplying by 0.85

Year 1: $0.85 \times £12\,500 = £10\,625$

Year 2: $0.85 \times £10\,625 = £9031.25$

Year 3: $0.85 \times £9031.25 = \mathbf{£7676.56}$

● This is the same as working out $(0.85)^3 \times £12\,500 = \mathbf{£7676.56}$

Compound Interest

Compound interest is where the bank pays interest on the interest already earned as well as on the original money.

Example
Becky has £3200 in her savings account and compound interest is paid at 3.2% per annum. How much will she have in her account after four years?

100% + 3.2% = 103.2%

= 1.032 ← This is the multiplier.

Year 1: 1.032 × £3200 = £3302.40

Year 2: 1.032 × £3302.40 = £3408.08

Year 3: 1.032 × £3408.08 = £3517.14

Year 4: 1.032 × £3517.14 = £3629.68

Total = **£3629.68**

A quicker way is to multiply £3200 by $(1.032)^4$

Number of years

£3200 × $(1.032)^4$ = **£3629.68**

Original   Multiplier

Simple Interest

Simple interest is the interest paid each year. It is the same amount each year.

The simple interest on £3200 invested for four years at 3.2% per annum would be:

$\frac{3.2}{100} \times 3200$ = £102.40 for one year

Interest over four years would be 4 × £102.40 = £409.60

Total in account after four years would be £3609.60

SUMMARY

● A quick way to work out repeated percentage change is to use a multiplier.

● Compound interest is where the bank pays interest on the interest already earned as well as on the original money.

● Simple interest is the interest paid each year. It is the same amount each year.

QUESTIONS

QUICK TEST

1. A car is bought for £8500. Two years later it is sold for £4105. Work out the percentage loss. Give your answer to 3 significant figures.

2. A flat was bought for £85 000 in 2013. The flat rose in value by 12% in 2014 and by 28% in 2015. How much was the flat worth at the end of 2015?

EXAM PRACTICE

1. Shamil invests £3000 in each of two bank accounts. The terms of the bank accounts are shown below.

Savvy Saver	Money Grows
Simple interest at 2.5% per annum.	Compound interest at 2.5% per annum.

Shamil says that he will earn the same amount of interest from both bank accounts in two years.

Decide whether Shamil is correct. You must show full working to justify your answer.

[3 marks]

Reverse Percentage Problems

In reverse percentage problems you are given the final amount after a **percentage increase** or **decrease**. You have to then find the value of the original quantity. These are quite tricky, so think carefully.

Example 1

The price of a television is reduced by 15% in the sales. It now costs £352.75
What was the original price?

- The sale price is 100% – 15% = 85% of the pre-sale price (x)

- 85% = 0.85 ← This is the **multiplier.**

- 0.85 × x = £352.75

$$x = \frac{£352.75}{0.85}$$

Original price was **£415**

Check:

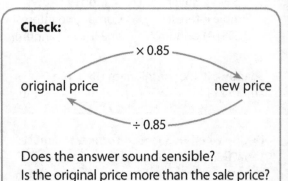

× 0.85

original price → new price

÷ 0.85

Does the answer sound sensible?
Is the original price more than the sale price?

Example 2

A mobile phone bill costs £169.20 including tax at 20%. What is the cost of the bill without the tax?

- The phone bill of £169.20 represents 100% + 20% = 120% of the original bill (x).

- 120% = 1.20 ← This is the **multiplier.**

- 1.2 × x = £169.20

$$x = \frac{£169.20}{1.2}$$

Original bill is **£141**

Check:

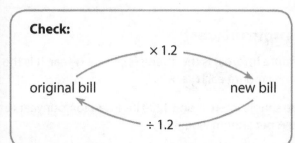

× 1.2

original bill → new bill

÷ 1.2

Example 3

The price of a washing machine is reduced by 5% in the sales. It now costs £323. What was the original price?

- The sale price is 100% – 5% = 95% of the pre-sale price (x).

- 95% = 0.95 ← This is the **multiplier**.

- 0.95 × x = £323

  $$x = \frac{£323}{0.95}$$

Original price was **£340**

Check:

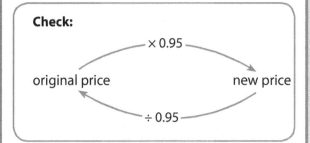

× 0.95

original price → new price

÷ 0.95

SUMMARY

- A reverse percentage problem is where you know the amount after a percentage change and want to find the original amount.

- Use a multiplier to work out reverse percentage problems.

- Always check that your answer seems sensible.

QUESTIONS

QUICK TEST

1. Each item listed below includes tax at 20%. Work out the original price of the item.

   a. A pair of shoes: £69

   b. A coat: £152.40

   c. A suit: £285

   d. A television: £525

EXAM PRACTICE

1. In a sale, normal prices are reduced by 12%. The sale price of a television is £220.

   Work out the normal price of the television.
   [3 marks]

2. Joseph says that the original price of a tablet device, which now costs £60 after a 15% reduction, was £70.59

   Is Joseph correct? Show your working.
   [3 marks]

Ratio and Proportion

Sharing a Quantity in a Given Ratio

A **ratio** is used to compare two or more related quantities. In **similar** shapes corresponding sides are in the same ratio.

The ratio $6:12$ can be simplified to $1:2$.

To share an amount in a given ratio, add up the individual parts and then divide the amount by this number to find one part.

> **Example**
> £155 is divided in the ratio of $2:3$ between Daisy and Tom. How much does each receive?
>
> $2 + 3 = 5$ parts ← Add up the total parts.
>
> 5 parts = £155
>
> 1 part = £155 ÷ 5 ← Work out what one part is worth.
>
> = £31
>
> So Daisy gets $2 \times £31 =$ **£62**
> and Tom gets $3 \times £31 =$ **£93**
>
> **Check:** £62 + £93
> = £155 ✔
>
>

Exchange Rates

Two quantities are in **direct proportion** when both quantities increase at the same rate.

> **Example**
> Samuel went on holiday to Spain. He changed £350 into euros. The exchange rate was £1 = €1.36 How many euros did Samuel receive?
>
> £1 = €1.36 so £350 = 350 × 1.36
>
> = **€476**

Best Buys

Use unit amounts to help you decide which is the better value for money.

> **Example**
> The same brand of breakfast cereal is sold in two different-sized packets.
>
> Which packet represents better value for money?
>
>
>
> Find the cost per gram for both boxes of cereal.
>
> 125 g costs £0.89 so $\frac{89}{125} = 0.712$p per gram
>
> 500 g costs £2.10 so $\frac{210}{500} = 0.42$p per gram
>
> **Since the larger box costs less per gram, it represents better value for money.**
>
> *N.B. There are many different ways of working out the answer to this question.*

Increasing and Decreasing in a Given Ratio

When increasing or decreasing in a given ratio, it is sometimes easier to find a unit amount.

Example

A recipe for four people needs 1600 g of flour. How much flour is needed to make the recipe for six people?

- Divide 1600 by 4, so 400 g for one person.
- Multiply by 6

So 6 × 400 g = **2400 g** of flour is needed for six people.

When two quantities are in **inverse proportion**, one quantity increases at the same rate as the other quantity decreases. For example, the time it takes to build a wall increases as the number of builders decreases.

It took four builders six days to build a wall.

Time for four builders is six days

Time for one builder is 6 × 4 = 24 days

It takes one builder four times as long to build the wall.

At the same rate it would take six builders $\frac{24}{6}$ = 4 days

Maps and Diagrams

Scales are often used on maps and diagrams. They are usually written as **ratios**.

Example

The scale on a road map is 1 : 25 000. Watford and St Albans are 60 cm apart on the map. Work out the real distance between them in km.

On a scale of 1 : 25 000, 1 cm on the map represents 25 000 cm on the ground.

60 cm represents 60 × 25 000 = 1 500 000 cm

1 500 000 ÷ 100 = 15 000 m ← Divide by 100 to change cm to m.

15 000 ÷ 1000 = 15 km ← Divide by 1000 to change m to km.

The distance between Watford and St Albans is **15 km**.

SUMMARY

- A ratio is used to compare two or more related quantities.

- Two quantities are in direct proportion when both quantities increase at the same rate.

- Two quantities are in inverse proportion when one quantity increases at the same rate as the other quantity decreases.

QUESTIONS

QUICK TEST

1. Divide £160 in the ratio 1 : 2 : 5

2. The cost of four ringbinders is £6.72
   Work out the cost of 21 ringbinders.

3. It took six builders four days to lay a patio. At the same rate how long would it take eight builders?

EXAM PRACTICE

1. Toothpaste is sold in three different-sized tubes.

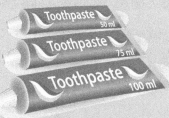

   50 ml is £1.24

   75 ml is £1.96

   100 ml is £2.42

   Which of the tubes of toothpaste is the better value for money? You must show full working in order to justify your answer. [3 marks]

2. Jessica buys a pair of jeans in England for £52. She then goes on holiday to America and sees an identical pair of jeans for $63. The exchange rate is £1 = $1.49.

   In which country are the jeans cheaper, and by how much? [2 marks]

Proportionality

Direct Proportion

In **direct proportion**, as one variable increases the other increases, and as one variable decreases the other decreases.

Example 1

If a is proportional to the square of b and $a = 5$ when $b = 4$, find the value of k (the **constant of proportionality**) and the value of a when $b = 8$.

● Change the sentence by adding the symbol ∝, which means 'is directly proportional to'.

$a \propto b^2$

● Replace ∝ with '$= k$' to make an equation.

$a = kb^2$

● Substitute the values given in the question in order to find k.

$5 = k \times 4^2$

Rearrange the equation.

$\frac{5}{16} = k$

● Replace k with the value just found.

$a = \frac{5}{16}b^2$

If $b = 8$ $a = \frac{5}{16} \times 8^2$

$a = \frac{5}{16} \times 64$

$a = \mathbf{20}$

Example 2

A train is accelerating out of a station at a constant rate.

The distance, d metres, travelled from the station varies directly as the square of the time taken, t seconds.

After 2.5 seconds the train has travelled 20 m.

a. Work out the formula connecting d and t.

$d \propto t^2$

$d = kt^2$

$20 = k \times 2.5^2$

$k = \frac{20}{2.5^2}$

$k = 3.2$

$d = \mathbf{3.2}t^2$

b. How long does the train take to travel 100 m?

$100 = 3.2t^2$

$t^2 = \frac{100}{3.2}$

$t^2 = 31.25$

$t = \sqrt{31.25}$

$t = \mathbf{5.59 \ seconds}$ (3 s.f.)

Inverse Proportion

In **inverse proportion**, as one variable increases the other decreases. If y is inversely proportional to x, then write this as:

$$y \propto \frac{1}{x} \text{ or } y = \frac{k}{x}$$

Example

p is inversely proportional to the cube of w. If $w = 2$ when $p = 5$, what is the value of w when $p = 10$? Give your answer to 3 decimal places.

$p \propto \dfrac{1}{w^3}$ ← Write the information with the proportionality sign.

$p = \dfrac{k}{w^3}$ ← Replace with the constant of proportionality.

$5 = \dfrac{k}{2^3}$

$5 \times 8 = k$

$k = 40$ ← Find the value of k.

$p = \dfrac{40}{w^3}$ ← Rewrite the formula.

$10 = \dfrac{40}{w^3}$ ← Find the value of w if $p = 10$

$w^3 = 4$

$w = \sqrt[3]{4}$

$w = \mathbf{1.587}$ (3 d.p.)

Rates of Change and Graphs

The **gradient** of a straight line is the rate of change.

$$\text{Gradient} = \frac{\text{change in } y}{\text{change in } x}$$

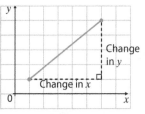

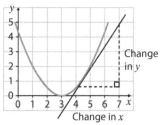

The gradient of a curved graph is only an estimate and is found by drawing a **tangent** at the point in which you are interested.

SUMMARY

● The notation ∝ means 'is directly proportional to'. This is often abbreviated to 'is proportional to' or 'varies as'.

● In direct proportion, if one variable increases the other variable also increases and if one variable decreases, the other variable also decreases.

● In inverse proportion, if one variable increases the other variable decreases.

QUESTIONS

QUICK TEST

1. Match these statements with the correct equation.

a. y is proportional to x	$y = \dfrac{k}{x}$
b. y is inversely proportional to the cube root of x	$y = kx^3$
c. y is inversely proportional to x	$y = kx$
d. y is proportional to x^3	$y = \dfrac{k}{\sqrt[3]{x}}$

EXAM PRACTICE

1. 🚫 a is directly proportional to $\sqrt{x}$. When $x = 4$, $a = 8$.

   What is the constant of proportionality and what is the value of x when $a = 64$? [4 marks]

2. y is inversely proportional to the square of x. When $x = 3$, $y = 10$

   a. Calculate y when $x = 2$ [4 marks]

   b. Calculate x when $y = 6$ [2 marks]

Similarity and Congruency 1

Objects that are exactly the same shape but different sizes are called **similar** shapes. One is an enlargement of the other.

Congruent shapes are identical to each other.

Congruent Triangles

Two triangles are congruent if any of the following sets of conditions is true.

S = side, **A** = angle, **R** = right angle, **H** = hypotenuse.

SSS: The three pairs of sides are equal.

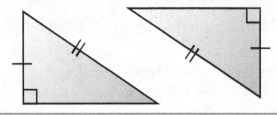

SAS: Two pairs of sides and the included angle are equal.

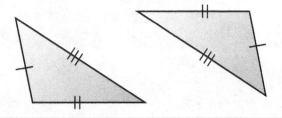

RHS: Each triangle contains a right angle. The hypotenuse and another pair of sides are equal.

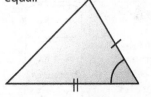

ASA or AAS: Two angles and a corresponding side are equal.

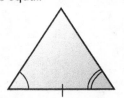

Proof

Congruency can be used to prove angle facts in an isosceles triangle.

An isosceles triangle has two equal sides. Here $AB = BC$

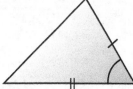

BD bisects the angle ABC therefore angle ABD = angle CBD

Side BD is common to both triangles.

Therefore SAS: triangles ABD and CBD are congruent.

So angle BAD = angle BCD; the base angles are equal, hence triangle BAC is an isosceles triangle.

Similarity

Questions about finding the missing lengths of similar figures are very common at GCSE. Remember that corresponding angles are equal. Corresponding lengths are in the same ratio.

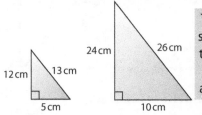

These triangles are similar – the sides of the second triangle are twice as long as those in the first.

Examples
1. Find the missing length labelled a.

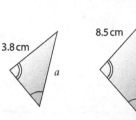

$$\frac{a}{12} = \frac{3.8}{8.5}$$

Corresponding lengths are in the same ratio.

$$a = \frac{3.8}{8.5} \times 12$$

Multiply both sides by 12.

$$a = \mathbf{5.36\,cm}\ (3\ \text{s.f.})$$

Examples (cont.)

2. Find the missing length labelled a in the diagram below.

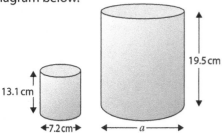

$$\frac{a}{7.2} = \frac{19.5}{13.1}$$

$$a = \frac{19.5}{13.1} \times 7.2 \quad \leftarrow \boxed{\text{Multiply both sides by 7.2}}$$

$$a = \textbf{10.7 cm} \text{ (3 s.f.)}$$

3. Calculate the missing length y in the right-angled triangle.

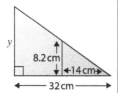

> First draw the individual triangles.

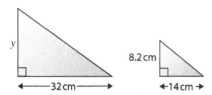

This gives an alternative way of writing the ratios to the method seen in the other examples. Both are correct!

$$\frac{y}{32} = \frac{8.2}{14} \quad \leftarrow \boxed{\begin{array}{c}\text{Write down the}\\\text{corresponding ratios.}\end{array}}$$

$$y = \frac{8.2}{14} \times 32 \quad \leftarrow \boxed{\text{Multiply both sides by 32.}}$$

$$y = \textbf{18.7 cm} \text{ (3 s.f.)}$$

SUMMARY

● **Congruent shapes are identical to each other.**

● **In similar shapes, corresponding angles are equal and corresponding lengths are in the same ratio.**

QUESTIONS

QUICK TEST

1. Calculate the lengths marked n in these similar shapes. Give your answers correct to 1 d.p.

a.

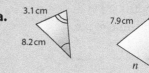

b.

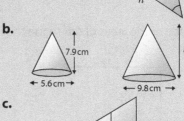

c.

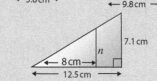

2. Are these two triangles congruent? Explain why.

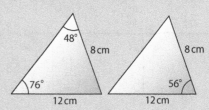

EXAM PRACTICE

1. The two triangles are similar. Work out the missing length x.　　　[3 marks]

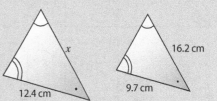

2. CDE is an equilateral triangle. F lies on DE. CF is perpendicular to DE.

Prove that triangle CFD is congruent to triangle CFE.

[3 marks]

Similarity and Congruency 2

Areas of Similar Figures

Areas of similar figures are not in the same ratio as their lengths.

If the corresponding lengths are in the ratio $a : b$, then their areas are in the ratio $a^2 : b^2$.

Examples

1. What is the ratio of the areas of the shapes below?

A = 1 cm²

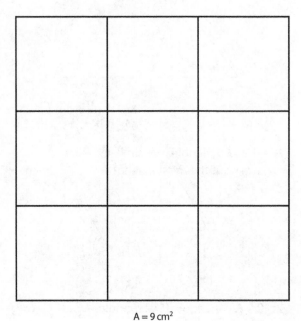

A = 9 cm²

The corresponding lengths of these squares are in the ratio 1 : 3 and their **areas are in the ratio 1 : 9**.

2. These two shapes are similar. The area of the larger shape, Q, is 81 cm². The area of the smaller shape, P, is 25 cm². If the height of the larger shape, Q, is 18.9 cm, work out the height of the smaller shape, P.

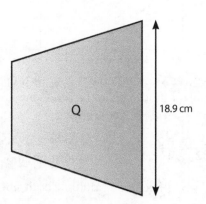

P x

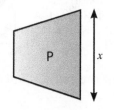

Q   18.9 cm

Area scale factor (k^2) is 25 : 81

Linear scale factor (k) is $\sqrt{25} : \sqrt{81} = 5 : 9$

Height of smaller shape is $\frac{5}{9} \times 18.9 = \textbf{10.5 cm}$

Volumes of Similar Figures

A similar result can be found when looking at the volumes of similar figures.

If the corresponding lengths are in the ratio $a : b$, then their volumes are in the ratio $a^3 : b^3$.

> For a scale factor k:
> The sides are k times bigger
> The areas are k^2 times bigger
> The volumes are k^3 times bigger

Example

These two solids are similar. If the volume of the smaller solid is 9 cm³, calculate the volume of the larger solid.

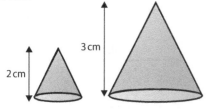

Linear scale factor is 2 : 3

Volume scale factor is $2^3 : 3^3 = 8 : 27$

Volume of larger solid is $\frac{27}{8} \times 9 = \textbf{30.375 cm}^3$

SUMMARY

● **To find the area of a similar shape:**

  – **If the corresponding lengths are in the ratio $a : b$, then their areas are in the ratio $a^2 : b^2$.**

● **To find the volume of a similar shape:**

  – **If the corresponding lengths are in the ratio $a : b$, then their volumes are in the ratio $a^3 : b^3$.**

QUESTIONS

QUICK TEST

1. The heights of two similar shapes are 8 cm and 12 cm. If the area of the larger shape is 64 cm², find the area of the smaller shape.

2. Two solids are similar. The heights of the two similar solids are 12 cm and 18 cm. The surface area of the smaller solid is 35 cm². Work out the surface area of the larger solid.

3. Cuboids A and B are similar. The volume of cuboid B is 140 cm³. Calculate the volume of cuboid A.

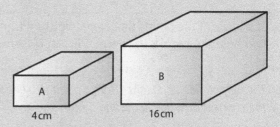

EXAM PRACTICE

1. 🚫 Two solids have the same density and are mathematically similar. The heights of the two solids are 8 cm and 24 cm. The smaller solid has a mass of 70 g.

   Work out the mass of the larger solid.   [3 marks]

2. Cuboids P and Q are similar. The surface area of cuboid P is 2160 cm². The surface area of cuboid Q is 60 cm². The volume of cuboid Q is 270 cm³.

   Calculate the volume of cuboid P.   [4 marks]

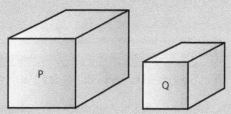

Loci

The **locus** of a point is the set of all the possible positions that the point can occupy, subject to some given condition or rule.

Types of Loci

<table>
<tr>
<td>

1. The locus of the points that are a constant distance from a fixed point is a circle.

</td>
<td>

Locus

• P

</td>
</tr>
<tr>
<td>

2. The locus of the points that are equidistant from two points X and Y is the perpendicular bisector of XY.

The perpendicular distance from a point to a line is the shortest distance to the line.

</td>
<td>

Perpendicular bisector

X ——————— Y

</td>
</tr>
<tr>
<td>

3. The locus of the points that are equidistant from two lines is the line that bisects the angle between the lines.

</td>
<td>

Locus

</td>
</tr>
<tr>
<td>

4. The locus of the points that are a constant distance from a line is a pair of lines parallel to the given line, one either side of it.

</td>
<td>

Locus

Locus

</td>
</tr>
<tr>
<td>

Sometimes you need to combine types 1 and 4.

A fixed distance from a line segment XY gives this locus.

</td>
<td>

X ——————— Y

Locus

</td>
</tr>
</table>

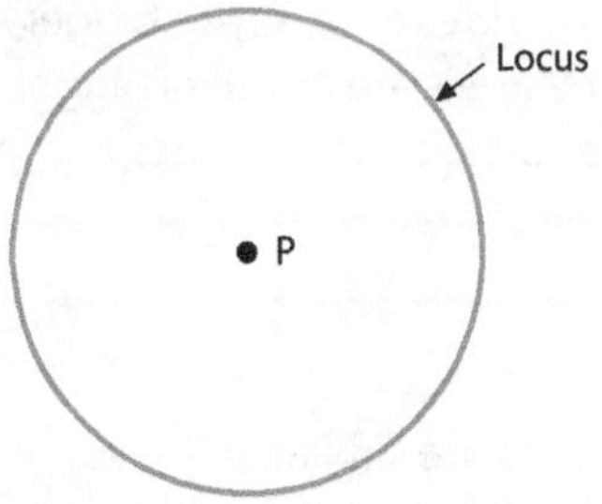

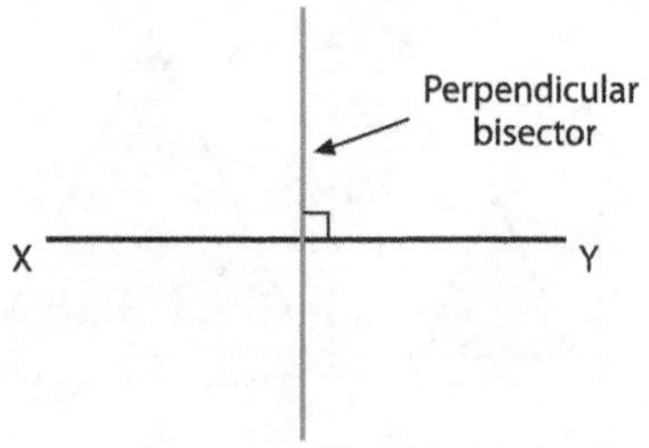

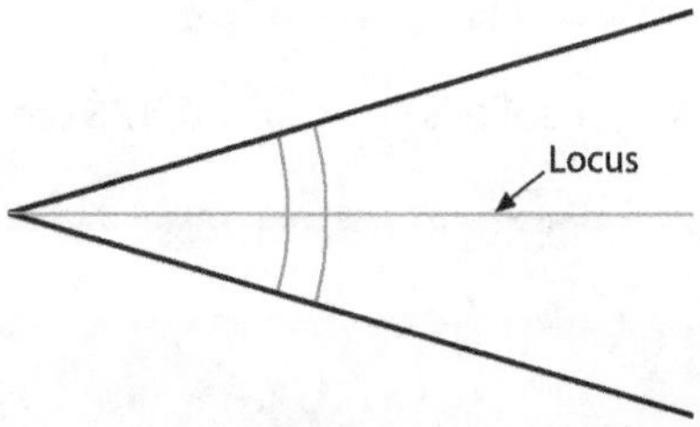

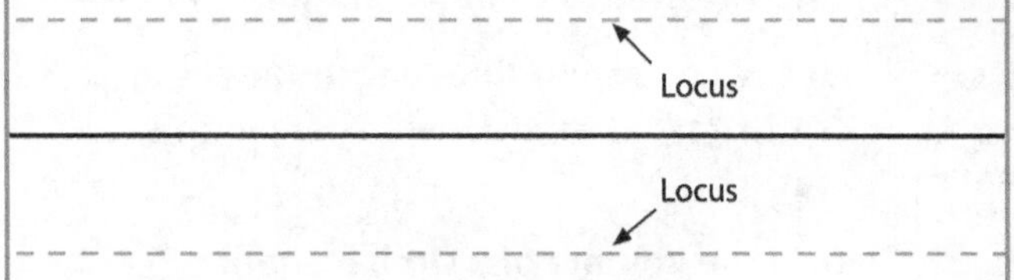

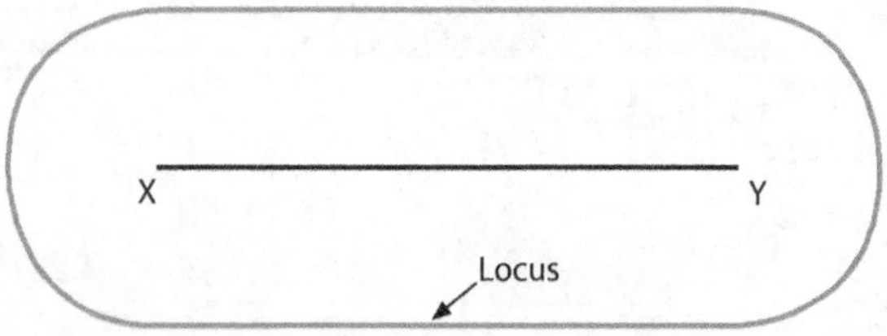

Example

Three radio transmitters form an equilateral triangle ABC with sides of 50 km. The range of the transmitter at A is 37.5 km, at B 30 km and at C 28 km. Using a scale of 1 cm to 10 km, construct a scale diagram to show where signals from all three transmitters can be received.

Below is a sketch not drawn to scale.

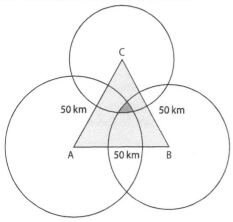

Please note that on your scale drawing the circle at A would have a radius of 3.75 cm. The circle at B would have a radius of 3 cm and the circle at C a radius of 2.8 cm.

The area where signals from all three transmitters can be received is shaded dark blue.

QUESTIONS

QUICK TEST

1. $ABCD$ is a rectangle.

   The rectangle is accurately drawn.

   Shade the set of points inside the rectangle which are more than 2 cm from point B and more than 1.5 cm from the line AD.

EXAM PRACTICE

1. The plan shows a garden drawn to a scale of 1 cm : 2 m. A and B are bushes and C is a pond.

   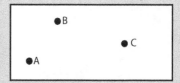

   A landscape gardener has decided:

   ● to lay a path right across the garden at an equal distance from each of the bushes, A and B.

   ● to lay a flower border around pond C at a distance of 2 m.

   Construct these features on the plan above.
   [2 marks]

Angles

Types of Angle

An angle is the amount of turning or rotation. Angles are measured in degrees. A circle is divided into 360 degrees.

- An **acute** angle is between 0° and 90°.
- An **obtuse** angle is between 90° and 180°.
- A **reflex** angle is between 180° and 360°.
- A **right angle** is 90°.

Angle Facts

Whenever lines meet or intersect, the angles they make follow certain rules:

Reading Angles

When asked to find angle ABC or $\angle ABC$ or $A\hat{B}C$, find the middle letter angle, i.e. at B:

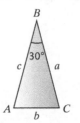

$\angle ABC = 30°$

When labelling a general triangle, the side opposite the vertex A is called a, the side opposite the vertex B is called b and the side opposite the vertex C is called c.

Angles on a straight line add up to 180°.	Angles in a triangle add up to 180°.	Angles in a quadrilateral add up to 360°.
$a + b + c = 180°$ 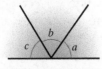	$a + b + c = 180°$	$a + b + c + d = 360°$
Angles at a point add up to 360°. $a + b + c + d = 360°$ 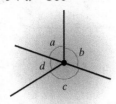	Vertically opposite angles are equal. $a = b, c = d$ $a + d = b + c = 180°$	An exterior angle of a triangle equals the sum of the two opposite interior angles. $c = a + b$  Since if $a + b + d = 180°$ (angles in a triangle add up to 180°) $d + c = 180°$ (angles on a straight line add up to 180°) Then $c = a + b$

Example
Find the missing angles.

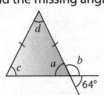

$a = \mathbf{64°}$ (opposite angles are equal)

$b = 180° - 64° = \mathbf{116°}$ (angles on a straight line add up to 180°)

$c = \mathbf{64°}$ (isosceles triangle base angles equal)

$d = \mathbf{52°}$ (angles in a triangle add up to 180°)

Parallel Lines

Three types of relationship are produced when a line called a transversal crosses a pair of parallel lines.

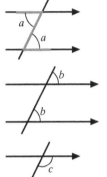

Alternate angles are equal.

Corresponding angles are equal.

Allied or **supplementary** angles add up to 180°. $c + d = 180°$

Angles in Polygons

There are two types of angle in a polygon – **interior** and **exterior**. A regular polygon has all sides and angles equal.

For a polygon with n sides:

● Sum of exterior angles = 360°
● Interior angle + exterior angle = 180°
● Sum of interior angles = $(n - 2) × 180°$ or $(2n - 4) × 90°$

For a regular polygon with n sides:

● Exterior angle = $\dfrac{360°}{n}$

Exterior angles

Interior angles

Example

A regular polygon has an interior angle of 108°. How many sides does it have?

180° – 108° = 72° (size of exterior angle)

360° ÷ 72° = **5 sides**

SUMMARY

● Angles add up to: 180° on a straight line, 360° at a point, 180° in a triangle, 360° in a quadrilateral.

● Alternate and corresponding angles are equal; supplementary angles add up to 180°.

● Polygons have interior and exterior angles.

QUESTIONS

QUICK TEST

1. Work out the size of the angles in the diagrams below.

   a.

   b.

   c.

2. Work out the size of the exterior angle of a 12-sided regular polygon.

EXAM PRACTICE

1. BCD is a straight line. Explain why BE and CF must be parallel. [2 marks]

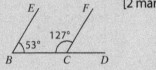

2. Work out the size of angle y in this polygon. [4 marks]

3. The diagram shows a regular octagon and a regular hexagon.

Find the size of the angle marked x. You must show all your working. [3 marks]

Translations and Reflections

Translations

Translations move figures from one position to another position. **Vectors** are used to describe the distance and direction of the translations.

A vector is written as $\begin{pmatrix} a \\ b \end{pmatrix}$.

a represents the horizontal distance and b represents the vertical distance.

The object and the image are **congruent** when the shape is translated, i.e. they are identical.

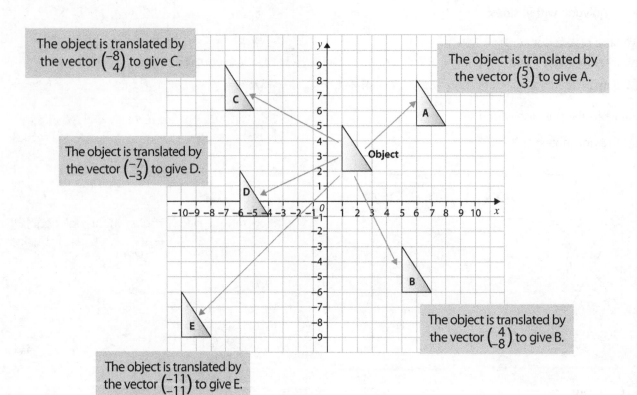

The object is translated by the vector $\begin{pmatrix} -8 \\ 4 \end{pmatrix}$ to give C.

The object is translated by the vector $\begin{pmatrix} 5 \\ 3 \end{pmatrix}$ to give A.

The object is translated by the vector $\begin{pmatrix} -7 \\ -3 \end{pmatrix}$ to give D.

The object is translated by the vector $\begin{pmatrix} 4 \\ -8 \end{pmatrix}$ to give B.

The object is translated by the vector $\begin{pmatrix} -11 \\ -11 \end{pmatrix}$ to give E.

Reflections

Reflections create an image of an object on the other side of the mirror line.

The mirror line is known as an axis of reflection.

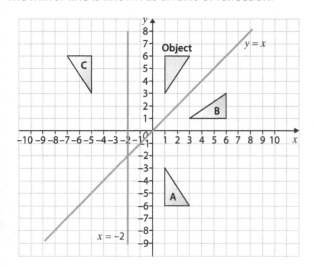

On the example above, the object is reflected in the x-axis (or $y = 0$) to give image A.

The object is reflected in the line $y = x$ to give image B.

The object is reflected in the line $x = -2$ to give image C.

The images and object are congruent.

QUICK TEST

1. For the diagram below, describe fully the transformation that maps:

   a. A onto B   **b.** B onto C

   c. A onto D   **d.** A onto E

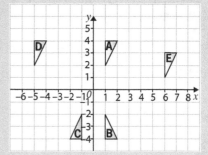

EXAM PRACTICE

1. Triangles A and B are shown on the grid below.

   a. Reflect triangle A in the line $x = 5$
   Label this image C.   [2 marks]

   b. Translate triangle B by the vector $\begin{pmatrix} 4 \\ 2 \end{pmatrix}$
   Label this image D.   [1 mark]

   c. Describe fully the single transformation which will map triangle A onto triangle B.
        [2 marks]

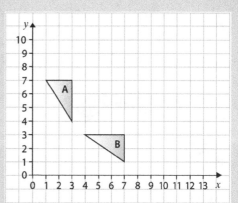

Rotation and Enlargement

Rotations

In a **rotation** the object is turned by a given angle about a fixed point called the **centre of rotation**. The size and shape of the figure are not changed, i.e. the image is **congruent** to the object.

On the example below, object A is rotated by 90° clockwise about (0, 0) to give image B.

Object A is rotated by 180° about (0, 0) to give image C.

Object A is rotated 90° anticlockwise about (−2, 2) to give image D.

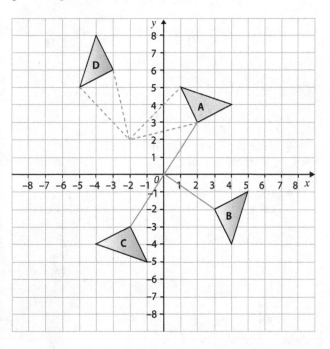

Enlargements

Enlargements change the size but not the shape of the object, i.e. the enlarged shape is **similar** to the object.

The **centre of enlargement** is the point from which the enlargement takes place.

The **scale factor** tells you what all lengths of the original figure have been multiplied by.

An enlargement with a scale factor between 0 and 1 makes the shape smaller.

Example
Describe fully the transformation that maps ABC onto $A'B'C'$.

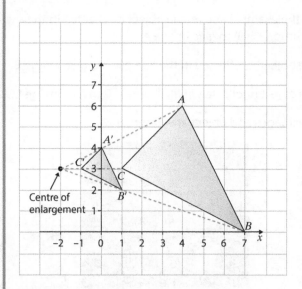

● The transformation is an enlargement.

● To find the centre of enlargement, join A to A', B to B', etc. and continue the line.

● Where all the lines meet is the centre of enlargement: (−2, 3).

● The length of $A'B'$ is a third of the length of AB.

The transformation is an enlargement by scale factor $\frac{1}{3}$. Centre of enlargement is (−2, 3).

Enlargements with a Negative Scale Factor

For an enlargement with a **negative scale factor**, the image is situated on the opposite side of the centre of enlargement.

On the example below, the triangle A (3, 4), B (9, 4) and C (3, 10) is enlarged with a scale factor of $-\frac{1}{3}$, with the centre of enlargement (0, 1). The enlargement is labelled $A'B'C'$.

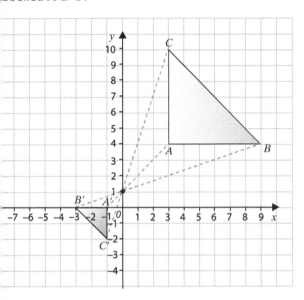

● $A'B'C'$ is on the opposite side of the centre of enlargement from ABC.

● Notice that the length of each side of the triangle $A'B'C'$ is one-third the size of the corresponding lengths in the triangle ABC.

SUMMARY

● **After a rotation, the image is congruent to the object.**

● **After an enlargement, the enlarged shape is similar to the object.**

● **An enlargement with a scale factor between 0 and 1 makes the shape smaller.**

● **An enlargement with a negative scale factor means that the image is on the opposite side of the centre of enlargement.**

QUESTIONS

QUICK TEST

1. Complete the diagram below, to show the enlargement of the shape by a scale factor of $\frac{1}{2}$. Centre of enlargement at (0, 0). Call the shape T.

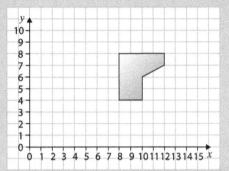

2. Rotate triangle ABC 90° anticlockwise about the point (0, 1). Call the triangle $A'B'C'$.

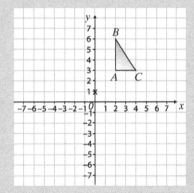

EXAM PRACTICE

1. The quadrilateral $ABCD$ is to be enlarged with a scale factor of $-\frac{1}{2}$ about the origin (0, 0). Draw the enlargement on the diagram. Call the shape $A'B'C'D'$. **[3 marks]**

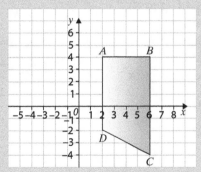

Circle Theorems

Circle Theorems to Know

1. The perpendicular bisector of any chord passes through the centre of the circle.

2. The angle in a semicircle is always 90°.

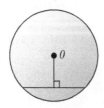

3. The radius and a tangent always meet at 90°.

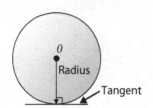

4. Angles in the same segment subtended by the same arc are equal, e.g. $A\hat{B}C = A\hat{D}C$.

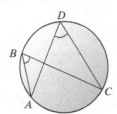

5. The angle at the centre is twice the angle at the circumference subtended by the same arc, e.g. $P\hat{O}Q = 2 \times P\hat{R}Q$.

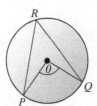

6. Opposite angles of a cyclic quadrilateral add up to 180°. (A cyclic quadrilateral is a four-sided shape with each vertex touching the circumference of the circle.)

i.e. $x + y = 180°$

$a + b = 180°$

7. The lengths of two tangents from a point are equal, i.e. $RS = RT$.

8. The angle between a tangent and a chord is equal to the angle in the alternate segment (that is the angle which is made at the edge of the circle by two lines drawn from the chord). This is known as the **Alternate Segment Theorem**.

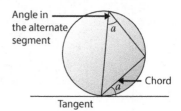

Angle in the alternate segment

Chord

Tangent

Example

Calculate the angles marked a–d in the diagram below. Give reasons for your answers.

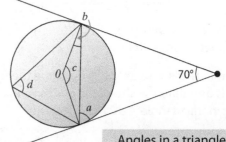

$a = \dfrac{180 - 70°}{2} = \mathbf{55°}$ ← Angles in a triangle add up to 180°. In an isosceles triangle, base angles are equal.

$b = 90° - 55° = \mathbf{35°}$ ← Radius and tangent meet at 90°.

$c = 180° - (2 \times 35°) = \mathbf{110°}$ ← Angles in a triangle add up to 180°.

$d = \dfrac{110°}{2} = \mathbf{55°}$ ← Angle at the centre is twice the angle at the circumference.

Proof

You might have to prove one of the circle theorems by writing out a set of logical steps.

> **Example**
> Prove that the angle at the centre of the circle is twice the angle at the circumference, subtended by the same arc.
>
>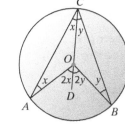
>
> Draw the line CO and extend it to D.
>
> $OA = OB = OC$, since they are all radii.
>
> Triangle OAC is isosceles, so angle OAC = angle $OCA = x$ (say)
>
> Triangle OBC is isosceles, so angle OBC = angle $OCB = y$ (say)
>
> Angle AOD = angle OAC + angle OCA (exterior angle of a triangle), i.e. angle $AOD = 2x$, similarly angle $BOD = 2y$
>
> Angle $AOB = 2x + 2y = 2(x + y) = 2 \times$ angle ACB
>
> Angle $AOB = 2 \times$ angle ACB

QUESTIONS

QUICK TEST

1. Some angles are written on cards. Match the missing angles in the diagrams below with the correct card. O is the centre of the circle.

$\boxed{53°}$ $\boxed{50°}$ $\boxed{62°}$ $\boxed{109°}$ $\boxed{126°}$

a.

b.

c.

d.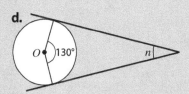

e.

EXAM PRACTICE

1. In the diagram, E, F and G are points on the circle, centre O.

   Angle $FGB = 71°$
   AB is a tangent to the circle at point G.

   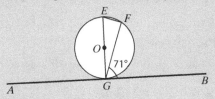

   a. Calculate the size of angle EGF. Give reasons for your answer. [2 marks]

   b. Calculate the size of angle GEF. Give reasons for your answer. [2 marks]

2. Prove that the angle subtended at the circumference by a semicircle equals 90°. [3 marks]

   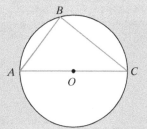

Pythagoras' Theorem

Pythagoras' Theorem states that 'For any right-angled triangle, the square on the **hypotenuse** is equal to the sum of the squares on the other two sides.'

$$a^2 + b^2 = c^2$$

Finding the Hypotenuse

Example
Find the hypotenuse in the right-angled triangle.

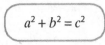

$n^2 = 7^2 + 12^2$ ← Square the two sides.

$n^2 = 49 + 144$ ← Add the two sides together.

$n^2 = 193$

$n = \sqrt{193}$ ← Square root.

$n = \textbf{13.9 cm}$ (3 s.f.) ← Round to 3 s.f.

On the non-calculator paper you would give this answer as $\sqrt{193}$ cm.

Finding a Short Side

Example
Find the length of p.

$15^2 = p^2 + 8^2$

$15^2 - 8^2 = p^2$

$225 - 64 = p^2$

$161 = p^2$

$\sqrt{161} = p$

$p = \textbf{12.7 cm}$ (3 s.f.)

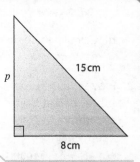

When finding a shorter length, remember to subtract.

Finding the Length of a Line Segment AB, Given the Coordinates of its End Points

Example
Find the length AB in this diagram.

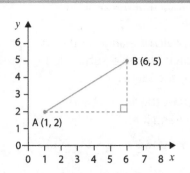

Horizontal distance $= 6 - 1 = 5$

Vertical distance $= 5 - 2 = 3$

$$AB^2 = 5^2 + 3^2$$

$$= 25 + 9$$

$$= 34$$

$$AB = \sqrt{34}$$

$$AB = \textbf{5.83 units} \text{ (3 s.f.)}$$

You could leave this as $\sqrt{34}$ units (**surd form**).

Solving a More Difficult Problem

Example
Calculate the vertical height of this isosceles triangle.

Using Pythagoras' Theorem gives:

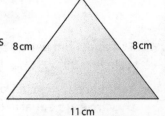

$$8^2 = h^2 + 5.5^2$$

$$64 = h^2 + 30.25$$

Split the triangle down the middle to make it right-angled.

$$64 - 30.25 = h^2$$

$$33.75 = h^2$$

$$\sqrt{33.75} = h$$

$$h = \textbf{5.81 cm} \text{ (3 s.f.)}$$

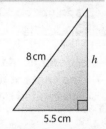

Pythagoras' Theorem in 3D

The length of the diagonal of a cuboid is found by applying Pythagoras' Theorem twice.

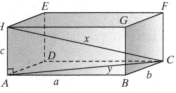

$x^2 = c^2 + y^2$ $y^2 = a^2 + b^2$ so $x^2 = a^2 + b^2 + c^2$

The length x of the longest diagonal in a cuboid with dimensions $a \times b \times c$ is:

$x^2 = a^2 + b^2 + c^2$

This is a 3D version of Pythagoras' Theorem.

Example

The cuboid has a length of 6 cm, a width of 4 cm and a height of 3 cm.

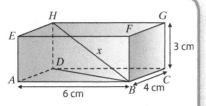

Work out the length of the longest diagonal, HB.

$x^2 = 4^2 + 6^2 + 3^2$

$x^2 = 16 + 36 + 9$

$x^2 = 61$

$x = \sqrt{61}$

$x = \textbf{7.81 cm}$ (2 d.p.)

SUMMARY

● **Pythagoras' Theorem states that 'For any right-angled triangle, the square on the hypotenuse is equal to the sum of the squares on the other two sides'.**

● **Remember:**

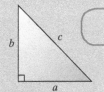

$$a^2 + b^2 = c^2$$

● **The length x of the longest diagonal in a cuboid with dimensions $a \times b \times c$ is**
$x^2 = a^2 + b^2 + c^2$

QUESTIONS

QUICK TEST

1. Work out the missing lengths labelled x in the diagrams below. Give your answers to 2 decimal places.

a.

b.

EXAM PRACTICE

1. Molly says: 'The angle x in this triangle is 90°.'

   Explain how Molly knows that without measuring the size of the angle.

   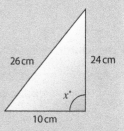

   [2 marks]

2. The diagram shows a room. Laminate flooring has been laid in the room. Laminate beading is now being placed along the walls of the room. Beading comes in 2.5 metre lengths and costs £1.74 per length.

   Calculate the cost of the beading for the room.

   [3 marks]

   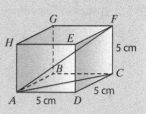

3. 🚫 Calculate the length of CD in this diagram. Leave your answer in surd form.   [2 marks]

   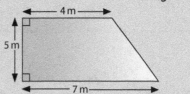

4. Here is a cube of side length 5 cm.

   Calculate the length of AF. Give your answer to 1 decimal place.

   [3 marks]

Trigonometry

Trigonometry in right-angled triangles can be used to find an unknown angle or length.

The sides of a right-angled triangle are given temporary names according to where they are in relation to a chosen angle θ.

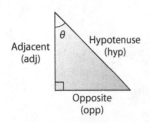

The trigonometric ratios are:

$$\text{Sine } \theta = \frac{\text{Opposite}}{\text{Hypotenuse}}$$

$$\text{Cosine } \theta = \frac{\text{Adjacent}}{\text{Hypotenuse}}$$

$$\text{Tangent } \theta = \frac{\text{Opposite}}{\text{Adjacent}}$$

Use **SOH – CAH – TOA** to remember the ratios.

Example: TOA means $\tan \theta = \dfrac{\text{opp}}{\text{adj}}$

Finding a Length

Example
Find the missing length y in the diagram.

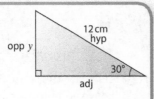

- Label the sides first.

- Decide on the ratio.

$$\sin 30° = \frac{\text{opp}}{\text{hyp}}$$

- Substitute in the values.

$$\sin 30° = \frac{y}{12}$$

$$12 \times \sin 30° = y$$

Multiply both sides by 12.

$$y = \mathbf{6\,cm}$$

Finding an Angle

Example
Calculate angle ABC.

Label the sides and decide on the ratio.

$$\cos \theta = \frac{\text{adj}}{\text{hyp}}$$

$$\cos \theta = \frac{15}{25}$$

$$\theta = \cos^{-1}\left(\frac{15}{25}\right)$$

To find the angle, you usually use the second function on your calculator.

$$= \mathbf{53.13°} \text{ (2 d.p.)}$$

It is important you know how to use your calculator when working out trigonometry questions. Check your calculator is set to work in degrees (not radians).

Angle of Elevation

The angle of elevation is the angle measured from the horizontal upwards.

Angle of Depression

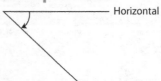

The angle of depression is the angle measured from the horizontal downwards.

Exact Values

The sine, cosine and tangent of some angles can be written exactly.

	0°	30°	45°	60°	90°
$\sin \theta$	0	$\frac{1}{2}$	$\frac{\sqrt{2}}{2}$	$\frac{\sqrt{3}}{2}$	1
$\cos \theta$	1	$\frac{\sqrt{3}}{2}$	$\frac{\sqrt{2}}{2}$	$\frac{1}{2}$	0
$\tan \theta$	0	$\frac{\sqrt{3}}{3}$	1	$\sqrt{3}$	

Example
Find the exact value of x in this triangle.

$\cos x = \dfrac{\text{adj}}{\text{hyp}}$ $\cos x = \dfrac{\sqrt{2}}{2}$ $x = \mathbf{45°}$

QUESTIONS

QUICK TEST

1. Work out the missing lengths labelled x in the diagrams below. Give your answers to 2 decimal places.

   a.

   b.
   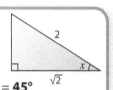

2. Work out the missing angles labelled x in the diagrams below. Give your answers to 1 decimal place.

   a.

   b.
   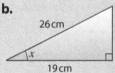

EXAM PRACTICE

1. The diagram represents a vertical mast, PN. The mast is supported by two metal cables, PA and PB, fixed to the horizontal ground at A and B.

   $BN = 12.6\,\text{m}$

   $PN = 19.7\,\text{m}$

   angle $PAN = 48°$

   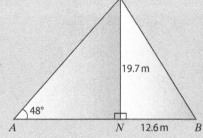

   a. Calculate the size of angle PBN. Give your answer correct to 3 significant figures.   [3 marks]

   b. Calculate the length of the metal cable PA. Give your answer correct to 3 significant figures.   [3 marks]

Trigonometry in 3D

The use of trigonometry and **Pythagoras' Theorem** can be applied when solving problems in three-dimensional figures.

Example

A, C and D are three points on horizontal ground.

A is 1600 m due west of D and C is 1420 m due south of D.

BD is a vertical tower.

The angle of elevation of B from C is 21°.

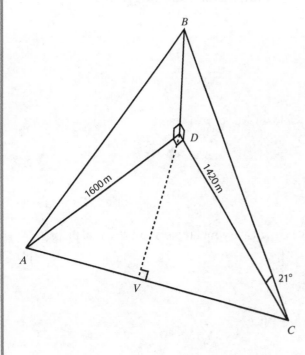

a. Calculate the height of the tower.

> Draw the triangle that you are working with.

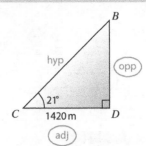

> Using triangle BCD.

$$\tan 21° = \frac{\text{opp}}{\text{adj}}$$

$$\tan 21° = \frac{BD}{1420}$$

$$1420 \times \tan 21° = BD$$

$$BD = \mathbf{545.1\ m}\ \text{(1 d.p.)}$$

b. Find the angle of elevation of B from A.

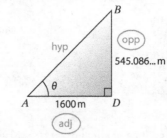

> Using triangle ABD.

$$\tan \theta = \frac{\text{opp}}{\text{adj}}$$

$$\tan \theta = \frac{545.086...}{1600}$$

$$\theta = \tan^{-1} 0.3406...$$

$$\theta = \mathbf{18.8°}\ \text{(1 d.p.)}$$

c. V is a point on AC which is nearest to D.
Calculate the angle of elevation of B from V.

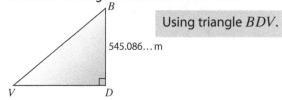

Using triangle BDV.

Since you only have one length in triangle BDV, you need to find some more information.

Using triangle ACD, find angle DCA.

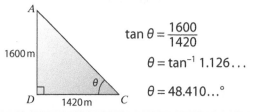

$$\tan \theta = \frac{1600}{1420}$$

$$\theta = \tan^{-1} 1.126\ldots$$

$$\theta = 48.410\ldots°$$

Now use triangle DVC to find length DV.

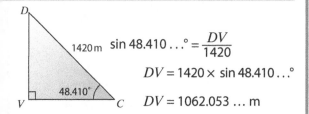

$$\sin 48.410\ldots° = \frac{DV}{1420}$$

$$DV = 1420 \times \sin 48.410\ldots°$$

$$DV = 1062.053\ldots \text{ m}$$

Now use triangle BDV to find angle BVD.

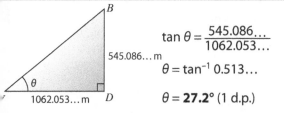

$$\tan \theta = \frac{545.086\ldots}{1062.053\ldots}$$

$$\theta = \tan^{-1} 0.513\ldots$$

$$\theta = \mathbf{27.2°} \text{ (1 d.p.)}$$

This is an example of a multi-step question, in that you have to do several parts to get to the answer. They are usually worth lots of marks.

SUMMARY

● When solving problems in 3D, it is useful to draw the right-angled triangle that is needed and apply Pythagoras' Theorem and trigonometry to find missing lengths and angles.

QUESTIONS

QUICK TEST

1. The diagram shows a cuboid.

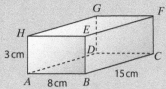

a. Are these statements true or false?

   i. The length HB is 8.54 cm (2 d.p.).

   ii. The length HC is 18 cm.

   iii. Angle ACD is 47°.

   iv. Angle HBA is 20.6° (1 d.p.).

b. Calculate the size of angle HCA to the nearest degree.

EXAM PRACTICE

1. Here is a square-based pyramid. Point E lies directly above M. M is the centre of the base.

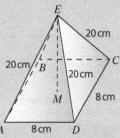

a. Calculate the height of the vertex E above the base. Give your answer to 2 decimal places.
                                     [4 marks]

b. Calculate the angle between ED and the base $ABCD$. Give your answer to 3 significant figures.
                                     [3 marks]

Sine and Cosine Rules

The sine and cosine rules allow you to solve problems in triangles that do not contain a right angle.

The Sine Rule

The standard way to write down the **sine** and **cosine** rules is to use the following notation for sides and angles in a general triangle.

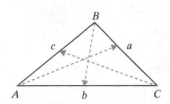

When finding a **length**:

$$\frac{a}{\sin A} = \frac{b}{\sin B} = \frac{c}{\sin C}$$

When finding an **angle**:

$$\frac{\sin A}{a} = \frac{\sin B}{b} = \frac{\sin C}{c}$$

Examples

1. Calculate the length of RS.

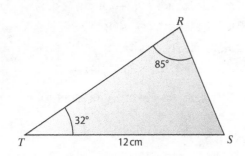

- Call RS, x, the length to be found.

- Since you have two angles and a side, use the sine rule:

$$\frac{x}{\sin 32°} = \frac{12}{\sin 85°}$$

- Rearrange to make x the subject:

$$x = \frac{12}{\sin 85°} \times \sin 32°$$

Length of $RS = $ **6.38 cm** (3 s.f.)

2. Calculate the size of angle CDE.

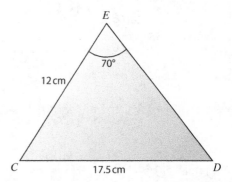

- Since you have two sides and an angle not enclosed by them, use the sine rule:

$$\frac{\sin D}{12} = \frac{\sin 70°}{17.5}$$

- Rearrange to give:

$$\sin D = \frac{\sin 70°}{17.5} \times 12$$

$$\sin D = 0.644\ldots$$

$$D = \sin^{-1} 0.644\ldots$$

$$D = \mathbf{40.1°} \text{ (3 s.f.)}$$

The Cosine Rule

When finding a **length**:

$$a^2 = b^2 + c^2 - (2bc \cos A)$$

When finding an **angle**:

$$\cos A = \frac{b^2 + c^2 - a^2}{2bc}$$

Examples

1. Calculate the length of JK.

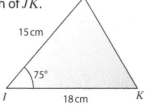

- Call JK, x, the length to be found.
- Since you have two sides and the included angle, use the cosine rule:

$$x^2 = 15^2 + 18^2 - (2 \times 15 \times 18 \times \cos 75°)$$

$$x^2 = 409.23\ldots$$

$$x = \sqrt{409.23\ldots}$$

$$JK = \mathbf{20.2\ cm} \text{ (3 s.f.)}$$

2. Calculate the size of angle x.

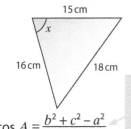

> If A = angle x then 18 cm is a, b = 16 cm and c = 15 cm.

$$\cos A = \frac{b^2 + c^2 - a^2}{2bc}$$

$$\cos x = \frac{16^2 + 15^2 - 18^2}{2 \times 16 \times 15}$$

> Substitute the values of a, b and c into the formula.

$$\cos x = \frac{256 + 225 - 324}{480}$$

$$\cos x = \frac{157}{480}$$

$$x = \cos^{-1} 0.32708\ldots$$

$$x = \mathbf{70.9°} \text{ (3 s.f.)}$$

QUESTIONS

QUICK TEST

1. For the questions below, decide whether the missing length x is correct.

a. $x = 11.49\ cm$

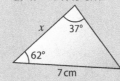

b. $x = 20.93\ cm$

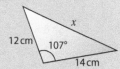

2. Calculate the missing angles in the diagrams below to the nearest degree:

a.

b.

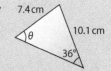

EXAM PRACTICE

1. The diagram shows a vertical pole AB on horizontal ground BCD.
BCD is a straight line.
D is 16 m from C.

The angle of elevation of A from C is 58°.
The angle of elevation of A from D is 32°.

Calculate the height of the pole. Give your answer correct to 3 significant figures. [5 marks]

Arc, Sector and Segment

Length of a Circular Arc

The length of a circular **arc** can be expressed as a fraction of the **circumference** of a circle.

Arc length $= \dfrac{\theta}{360°} \times 2\pi r$

where θ is the angle subtended at the centre.

Example
Work out the length of the minor arc AB.

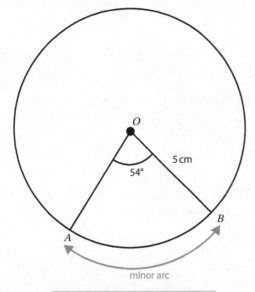

minor arc

O is the centre of the circle.

$= \dfrac{54°}{360°} \times 2 \times \pi \times 5$

$= \dfrac{3}{2}\boldsymbol{\pi}$ **cm** (non-calculator paper) or

4.71 cm (2 d.p.) (calculator paper)

Area of a Sector

The **area** of a **sector** can be expressed as a fraction of the area of a circle.

Area of a sector $= \dfrac{\theta}{360°} \times \pi r^2$

where θ is the angle subtended at the centre.

Example
Work out the area of the minor sector AOB.

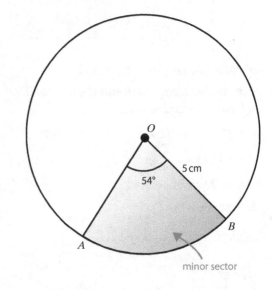

minor sector

$= \dfrac{54°}{360°} \times \pi \times 5^2$

$= \dfrac{15}{4}\boldsymbol{\pi}$ **cm²** (non-calculator paper) or

11.78 cm² (2 d.p.) (calculator paper)

Area of a General Triangle

If you know the length of two sides of a triangle and the included angle, you can find the area.

Area = $\frac{1}{2} \times a \times b \times \sin C$

two lengths ⟶ ⟵ included angle

Example

Work out the area of the triangle below.

Area = $\frac{1}{2} \times 10 \times 12 \times \sin 58°$

= **50.88 cm²** (2 d.p.)

10 cm
58°
12 cm

Area of a Segment

The area of a **segment** can be worked out in two stages:

1. Calculate the area of the sector and the area of the triangle.

2. Subtract the area of the triangle from the area of the sector.

Example

Work out the area of the segment.

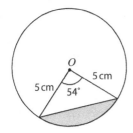

O
5 cm
5 cm
54°

Segment area

= $\left(\frac{54°}{360°} \times \pi \times 5^2\right) - \left(\frac{1}{2} \times 5 \times 5 \times \sin 54°\right)$

= 11.78... − 10.11...

= **1.67 cm²** (3 s.f.)

QUESTIONS

QUICK TEST

1. For this diagram, calculate:

   a. the arc length (x)

   b. the sector area

   c. the area of the shaded segment

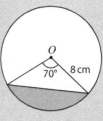

O 115°
7 cm
x

2. Decide whether the statement below is true or false. You must show sufficient working out to justify your answer.

   The area of the shaded segment is 9.03 cm² (3 s.f.).

O
70° 8 cm

EXAM PRACTICE

1. The diagram shows a sector of a circle, AOB, centre O, radius 6 cm.

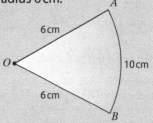

A
6 cm
O
10 cm
6 cm
B

The arc length of the sector is 10 cm.

Calculate the area of the sector in cm². [4 marks]

Surface Area and Volume

A **prism** is any solid that can be cut up into slices that are all the same shape. This is known as having a uniform **cross-section**.

Key Formulae

● **Volume of a Prism**

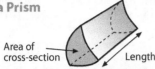

Area of cross-section — Length

> Volume = area of cross-section × length
> $$V = A \times l$$

● **Sphere**

> Volume of a sphere $= \frac{4}{3}\pi r^3$

> Surface area of a sphere $= 4\pi r^2$

● **Pyramids and Cones**

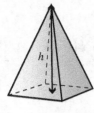

A cone is simply a pyramid with a circular base.

> Volume of a pyramid $= \frac{1}{3} \times$ area of base × height

> Volume of a cone $= \frac{1}{3} \times \pi \times r^2 \times$ height
> $$V = \frac{1}{3}\pi r^2 h$$

> Curved surface area of cone $= \pi r l$ (l is the slant height)

Examples

1. Find the volume of this cylinder:

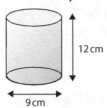

12 cm

9 cm

$$V = \pi r^2 h$$
$$= \pi \times 4.5^2 \times 12$$
$$= \mathbf{243\pi \ cm^3}$$

> This answer is left in terms of π.

2. The diagram below shows a cone.

12 cm

5 cm

a. Calculate the volume of the cone. Give your answer to 2 decimal places.

$$V = \frac{1}{3}\pi r^2 h$$
$$= \frac{1}{3} \times \pi \times 5^2 \times 12$$
$$= 100\,\pi$$
$$= \mathbf{314.16 \ cm^3} \text{ (2 d.p.)}$$

b. Calculate the total surface area of the cone. Give your answer to 2 decimal places.

$$\text{Total surface area} = \text{curved surface area} + \text{area of circle}$$

Slant height, $l = \sqrt{12^2 + 5^2}$
$$= \sqrt{144 + 25}$$
$$= \sqrt{169}$$
$$= 13 \text{ cm}$$

$$A = \pi r l + \pi r^2$$
$$= \pi \times 5 \times 13 + \pi \times 5^2$$
$$= 65\pi + 25\pi$$
$$= 90\pi$$
$$= \mathbf{282.74 \ cm^2} \text{ (2 d.p.)}$$

Example

The diagram shows a solid toy. The mass of the toy is 637 grams.

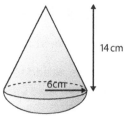

Work out the density of the toy. Give your answer to 2 decimal places.

First, work out the volume of the toy.

Volume of cone $= \frac{1}{3}\pi r^2 h$

$= \frac{1}{3} \times \pi \times 6^2 \times 14$

$= 527.787\ldots$ cm^3

Volume of sphere $= \frac{4}{3}\pi r^3$

Volume of sphere $= \frac{4}{3} \times \pi \times 6^3$

Volume of hemisphere $= \frac{904.7786\ldots}{2}$

$= 452.389\ldots$ cm^3

Total volume $= 527.787\ldots + 452.389\ldots$

$= 980.176\ldots$ cm^3

Density $= \frac{\text{mass}}{\text{volume}}$

Density $= \frac{637}{980.176\ldots}$

Density $= \textbf{0.65 g/cm}^3$ (2 d.p.)

SUMMARY

- **Volume of a prism = area of cross-section × length**

- **Volume of a sphere $= \frac{4}{3}\pi r^3$**

- **Surface area of a sphere $= 4\pi r^2$**

- **Volume of a pyramid $= \frac{1}{3} \times$ area of base × height**

- **Volume of a cone $= \frac{1}{3}\pi r^2 h$**

- **Curved surface area of cone $= \pi r l$**
  (l is the slant height)

QUESTIONS

QUICK TEST

1. The volumes of the solids below have been calculated. Match the correct solid with the correct volume.

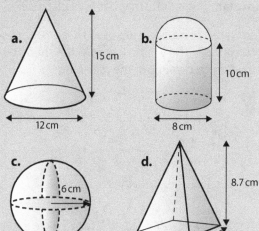

905 cm^3 245 cm^3 565 cm^3 637 cm^3

EXAM PRACTICE

1. A cone has a volume of 15 m^3. The vertical height of the cone is 2.1 m.

    Calculate the radius of the base of the cone. Give your answer to 3 significant figures.   [4 marks]

2.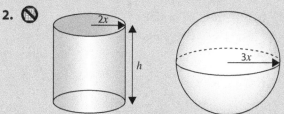

    The diagrams show a solid cylinder and a solid sphere. The measurements are in cm.

    The cylinder has a radius of $2x$ cm and a height, h cm. The sphere has a radius of $3x$ cm. The volume of the cylinder is equal to the volume of the sphere.

    Find an expression for h in terms of x.   [4 marks]

Vectors

Representing Vectors

A **vector** is a quantity that has both distance and direction.

A **scalar** is a quantity that has only distance.

Four types of notation are used to represent vectors. For example, the vector shown in the shape below can be referred to as any of the following:

$\begin{pmatrix} 5 \\ 2 \end{pmatrix}$ <u>a</u> $\overrightarrow{AB}$ **a**

The direction of the vector is usually shown by an arrow. On the diagram above, the vector $\overrightarrow{AB}$ is shown by an arrow.

● If $\overrightarrow{DE} = k\overrightarrow{AB}$, then $\overrightarrow{AB}$ and $\overrightarrow{DE}$ are parallel and the length of $\overrightarrow{DE}$ is k times the length of $\overrightarrow{AB}$.

$$\overrightarrow{AB} = \begin{pmatrix} 5 \\ 2 \end{pmatrix}$$

$$\overrightarrow{DE} = \begin{pmatrix} 10 \\ 4 \end{pmatrix} = 2\begin{pmatrix} 5 \\ 2 \end{pmatrix}$$

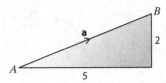

$$\overrightarrow{DE} = 2\overrightarrow{AB}$$

● If two vectors are equal, they are parallel and equal in length.

$\overrightarrow{AB}$ is equal to **a**.

● If the vector $\mathbf{c} = \begin{pmatrix} 3 \\ 4 \end{pmatrix}$ then the vector $-\mathbf{c}$ is in the opposite direction to **c**.

$$-\mathbf{c} = \begin{pmatrix} -3 \\ -4 \end{pmatrix}$$

Addition and Subtraction of Vectors

The **resultant** of two vectors is found by adding them.

Vectors must always be added end to end so that the arrows follow on from each other. A resultant is usually labelled with a double arrow.

Addition

The triangle shows the triangle law of vector addition.

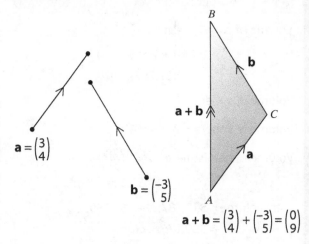

$$\mathbf{a} + \mathbf{b} = \begin{pmatrix} 3 \\ 4 \end{pmatrix} + \begin{pmatrix} -3 \\ 5 \end{pmatrix} = \begin{pmatrix} 0 \\ 9 \end{pmatrix}$$

To take the route directly from A to B is equivalent to travelling via C. Hence we can represent $\overrightarrow{AB}$ as **a** + **b**.

Subtraction

Vectors can also be subtracted, for example:

a – **b** can be interpreted as **a** + (–**b**).

$$\mathbf{a} + (-\mathbf{b}) = \begin{pmatrix} 3 \\ 4 \end{pmatrix} + \begin{pmatrix} 3 \\ -5 \end{pmatrix}$$

$$\therefore \ \mathbf{a} - \mathbf{b} = \begin{pmatrix} 6 \\ -1 \end{pmatrix}$$

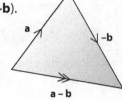

Position Vectors

The position vector of a point B is the vector $\overrightarrow{OB}$, where O is the origin.

In the diagram, the position vectors of B and C are **b** and **c** respectively. Using this notation:

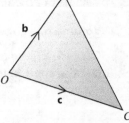

$$\overrightarrow{BC} = -\mathbf{b} + \mathbf{c} = \mathbf{c} - \mathbf{b}$$

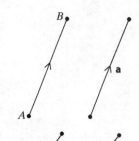

Example

OAB is a triangle. Given that $\vec{OA} = \mathbf{a}$, $\vec{OB} = \mathbf{b}$ and that N splits $\vec{AB}$ in the ratio 1 : 2, prove that $\vec{ON} = \frac{1}{3}(2\mathbf{a} + \mathbf{b})$

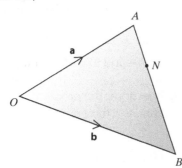

Find $\vec{AB}$ first:

$\vec{AB} = \vec{AO} + \vec{OB}$ | Go from A to B via O. |

    $= -\mathbf{a} + \mathbf{b}$

$\vec{ON} = \vec{OA} + \vec{AN}$ | $\vec{AN} = \frac{1}{3}\vec{AB}$ |

    $= \mathbf{a} + \frac{1}{3}(-\mathbf{a} + \mathbf{b})$

    $= \mathbf{a} - \frac{1}{3}\mathbf{a} + \frac{1}{3}\mathbf{b}$

    $= \frac{2}{3}\mathbf{a} + \frac{1}{3}\mathbf{b}$

$\vec{ON} = \frac{1}{3}(2\mathbf{a} + \mathbf{b})$

SUMMARY

● **The direction of a vector is shown with an arrow.**

● **The resultant of two vectors is found by vector addition or subtraction.**

● **Vectors must always be combined end to end.**

QUICK TEST

1. $OABC$ is a parallelogram.
   AB is parallel to OC.
   OA is parallel to CB.

   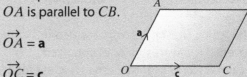

   $\vec{OA} = \mathbf{a}$

   $\vec{OC} = \mathbf{c}$

   a. Express in terms of $\mathbf{a}$ and $\mathbf{c}$:

      i. $\vec{AC}$

      ii. $\vec{BO}$

   b. N is the midpoint of $\vec{AC}$:

      Express $\vec{ON}$ in terms of $\mathbf{a}$ and $\mathbf{c}$.

EXAM PRACTICE

1. $ABCDEF$ is a regular hexagon.

   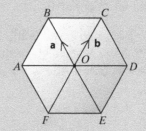

   Given that $\vec{OB} = \mathbf{a}$ and $\vec{OC} = \mathbf{b}$:

   a. Find in terms of $\mathbf{a}$ and $\mathbf{b}$ the vectors:

      i. $\vec{BC}$ [1 mark]

      ii. $\vec{AD}$ [1 mark]

   b. Write down the vector $\vec{FE}$.   [1 mark]

   c. What geometrical fact is exhibited by the vectors $\vec{FE}$ and $\vec{AD}$?   [2 marks]

Probability

Probability helps you predict the outcome of an event.

> The expected number of outcomes = number of trials × probability

Tree diagrams show the possible outcomes of two or more events. There are two rules you need to know. These rules also work for more than two events.

● The OR rule

If two events are **mutually exclusive**, the probability of A or B happening is found by adding the probabilities.

> P(A or B) = P(A) + P(B)

● The AND rule

If two events are **independent**, the probability of A and B happening together is found by multiplying the separate probabilities.

> P(A and B) = P(A) × P(B)

In probability questions, the product rule for counting can be used to find the total number of combinations possible.

For example, if there are A ways of doing task 1 and B ways of doing task 2, there are $A \times B$ ways of doing both tasks.

Example

A bag contains three red and four blue counters. A counter is taken from the bag at random, its colour is noted and then it is replaced in the bag. A second counter is then taken out of the bag.

Draw a tree diagram to illustrate this information. Remember that the probabilities on the branches leaving each point on the tree add up to 1.

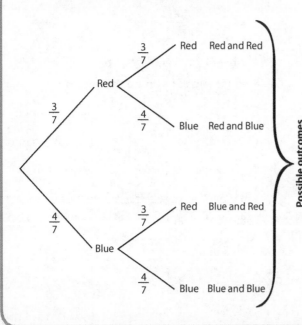

Possible outcomes

Work out the probability of:

a. picking two blues

> To find the probability of picking a blue AND a blue, multiply along the branches.

$$P(B) \times P(B)$$
$$= \frac{4}{7} \times \frac{4}{7} = \frac{16}{49}$$

b. picking one of either colour

> The probability of picking one of either colour is the probability of picking a blue and a red OR a red and a blue. You need to use the AND rule and the OR rule.

P(blue and red)
$$= P(B) \times P(R)$$
$$= \frac{4}{7} \times \frac{3}{7} = \frac{12}{49} \quad \leftarrow \text{The AND rule}$$

P(red and blue)
$$= P(R) \times P(B)$$
$$= \frac{3}{7} \times \frac{4}{7} = \frac{12}{49} \quad \leftarrow \text{The AND rule}$$

P(one of either colour)
$$= \frac{12}{49} + \frac{12}{49} = \frac{24}{49} \quad \leftarrow \text{The OR rule}$$

If the counters in the example had not been replaced, it would be an example of **conditional probability**. This is when the outcome of the second event is dependent on the outcome of the first. The tree diagram for the example without replacement would look like this:

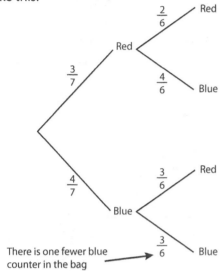

There is one fewer blue counter in the bag

Probability of picking two blues:

$$P(B) \times P(B)$$
$$= \frac{4}{7} \times \frac{3}{6} = \frac{12}{42}$$
$$= \frac{2}{7}$$

SUMMARY

- **If two events are mutually exclusive, the probability of A or B happening is found by adding the probabilities.**

  $$P(A \text{ or } B) = P(A) + P(B)$$

- **If two events are independent, the probability of A and B happening is found by multiplying the probabilities.**

  $$P(A \text{ and } B) = P(A) \times P(B)$$

- **These rules also work for more than two events.**
- **Remember that the probabilities on the branches leaving each point on a tree diagram should add up to 1.**

QUICK TEST

1. Sangeeta has a biased dice. The probability of getting a three is 0.4. She rolls the dice twice.

   a. Complete the tree diagram.

   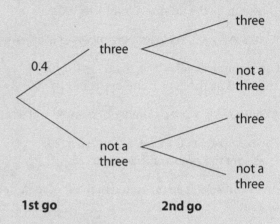

   1st go   **2nd go**

   b. Work out the probability that she gets:

   i. two threes

   ii. exactly one three

EXAM PRACTICE

1. A bag contains three red, four blue and two green beads. A bead is picked out of the bag at random and its colour noted. It is **not** replaced in the bag. A second bead is picked out at random.

   Work out the probability that two different-coloured beads are chosen.   [4 marks]

2. On her way to work, Mrs Harris drives down a country lane. The probability that she meets a tractor in the lane is 0.2. If she meets a tractor, the probability that she is late for work is 0.6. If she does not meet a tractor, the probability that she is late for work is 0.1.

   a. What is the probability that Mrs Harris meets a tractor and is late for work?   [2 marks]

   b. What is the probability that Mrs Harris is not late for work?   [3 marks]

Sets and Venn Diagrams

Set Notation

- A **set** is a collection of objects, which are called the elements or members of the set.

  For example: Set A = {1, 2, 3, ...100}
  This set is all the numbers from 1 to 100.

- A **subset** is a set made from members of a larger set.

  For example: Set B = {2, 4, 6, ...100}
  This set is all the even numbers from 1 to 100.

- A **finite set** is a given number of members of a set.

  For example: Set C = {2, 4, 6, ...20}
  These are the even numbers up to 20.

- An **infinite set** is an unlimited number of members of a set.

  For example: Set D = {2, 4, 6, ...}
  These are all the even numbers.

- An **empty** or **null set** Ø is a set that contains no members.

- A **universal set** is a set that contains all possible members.

Union of Sets (∪)

Set A ∪ B contains members belonging to A or B or both.

> **Example**
> If A = {2, 3, 4} and B = {4, 5, 6, 7}, then A ∪ B = {2, 3, 4, 5, 6, 7}.

Intersection of Sets (∩)

Set A ∩ B contains members belonging to both A and B.

> **Example**
> If A = {2, 3, 4} and B = {1, 3, 5}, then A ∩ B = {3}, since this is the only number in both sets.

If A and B have no members in common, then A ∩ B = Ø, i.e. the empty set.

Venn Diagrams

Relationships between sets can be shown on a **Venn diagram**. The universal set contains all the elements being discussed and is shown as a rectangle.

Examples

1. A + A′ = ξ

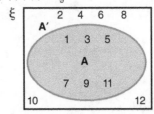

A = {1, 3, 5, 7, 9, 11}

A′ = {2, 4, 6, 8, 10, 12} ← *A′ is the set of numbers **not** in set A.*

ξ = {1, 2, 3, ...12}

2. A ∪ B

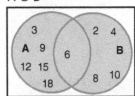

A = {3, 6, 9, 12, 15, 18}

B = {2, 4, 6, 8, 10}

A ∪ B = {2, 3, 4, 6, 8, 9, 10, 12, 15, 18}

3. A ∩ B

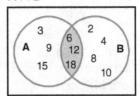

A = {3, 6, 9, 12, 15, 18}
These are multiples of 3.

B = {2, 4, 6, 8, 10, 12, 18}
These are multiples of 2.

A ∩ B = {6, 12, 18}
These are multiples of 6.

Using Venn Diagrams to Solve Probability Questions

Venn diagrams can be used to help solve probability questions.

Example

Out of 40 students, 14 are taking French and 29 are taking German.

a. If five students are in both classes, how many students are in neither class?

b. How many are in at least one class?

c. What is the probability that a randomly-chosen student is only taking German?

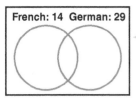

First draw the universal set for the 40 students, with two overlapping circles labelled with the total for each.

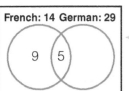

Since five students are taking both classes, put "5" in the overlap. Five of the 14 French students have been accounted for, leaving nine students taking French but not German, so put "9" in the "French only" part of the "French" circle.

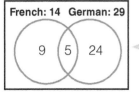

Five of the 29 German students have been accounted for, leaving 24 students taking German but not French, so put "24" in the "German only" part of the "German" circle.

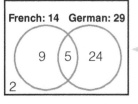

This tells you that a total of 9 + 5 + 24 = 38 students are taking either French or German (or both). This leaves two students unaccounted for, so they must be the ones taking neither class.

From this Venn diagram, the answers are:

a. **Two** students are in neither class.

b. There are **38** students in at least one of the classes.

c. There is a $\frac{24}{40} = 0.6 = 60\%$ probability that a randomly-chosen student in this group is taking only German.

SUMMARY

- Set A ∪ B contains members belonging to A or B or both.

- Set A ∩ B contains members belonging to both A and B.

- If A and B have no members in common, then A ∩ B = Ø, i.e. the empty set

- An empty or null set Ø contains no members.

- A universal set contains all possible members.

- Sets can be shown in a Venn diagram.

QUESTIONS

QUICK TEST

1. If C = {2, 3, 4, 5, 6, 7} and D = {6, 7, 8, 9} then write down:

    a. C ∪ D   **b.** C ∩ D

EXAM PRACTICE

1. Alex asked 50 people which type of chocolate they liked from plain (P), milk (M) and white (W). All 50 people liked at least one of the types.
19 people liked all three flavours.
16 people liked plain and milk chocolate but did not like white chocolate.
21 people liked milk and white chocolate.
24 people liked plain and white chocolate.
40 people liked milk chocolate.
1 person liked only white chocolate.

    a. Draw a Venn diagram to represent this information. [3 marks]

    Alex chose at random one of the 50 people.

    b. Work out the probability that the person liked plain chocolate. [2 marks]

    c. Given that the person selected at random liked plain chocolate, find the probability that this person liked exactly one other type of chocolate. [2 marks]

Statistical Diagrams

Diagrams to Compare Data

Pictograms, pie charts and vertical line graphs can display data. A time-series graph can show how data changes over time.

Dual bar charts can be used to compare data. In a dual bar chart, two (or more) bars are drawn side by side.

Example

Thomas has carried out a survey on some students' favourite sport. Here are his results:

Favourite sport	No. of boys	No. of girls
Swimming	6	8
Football	15	3
Hockey	5	10
Tennis	9	14

Draw a dual bar chart of these results.

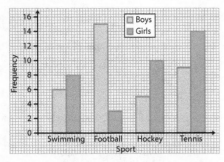

A compound bar chart also helps to compare two or more sets of data, e.g. for Thomas's results:

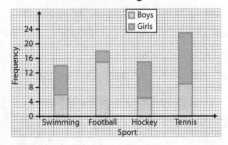

Both bar charts above show that considerably more boys than girls like football and considerably more girls than boys like tennis.

Scatter Diagrams and Correlation

Scatter diagrams are used to show two sets of data at the same time. They are important because they show the **correlation** (connection) between the sets.

Positive Correlation

Both variables are increasing.

For example, the taller you are, the more you are likely to weigh.

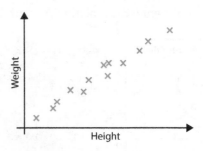

Negative Correlation

As one variable increases, the other decreases.

For example, as the temperature increases, the sales of woollen hats are likely to decrease.

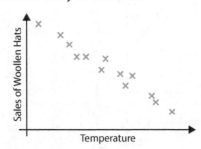

Zero Correlation

Little or no linear relationship between the variables.

For example, there is no connection between your height and your mathematical ability.

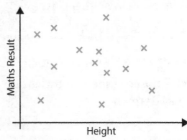

Line of Best Fit

The **line of best fit** goes as close as possible to all the points. There is roughly an equal number of points above the line and below it.

The scatter diagram below shows the Science and Maths percentages scored by some students.

- The line of best fit goes in the direction of the data.
- The line of best fit can be used to estimate results.
- You can estimate that a student with a Science percentage of 30 would get a Maths percentage of about 18.
- You can estimate that a student with a Maths percentage of 50 would get a Science percentage of about 54.

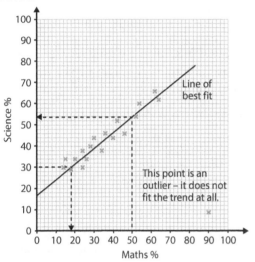

SUMMARY

- Dual bar charts have two or more bars drawn side by side. They can be used to compare data.

- Positive correlation: both variables are increasing.

- Negative correlation: as one variable increases, the other decreases.

- Zero correlation: little or no linear correlation between the variables.

- A line of best fit should be as close as possible to all the points. It is in the direction of the data.

QUESTIONS

QUICK TEST

1. Audrey owns a flower shop. This bar chart compares the percentage of types of flowers sold over the last two years.

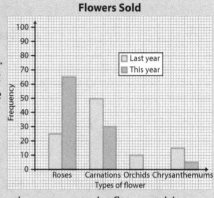

   a. Write down the most popular flower sold:
      i. this year   ii. last year
   b. Work out the percentage of chrysanthemums sold: i. this year   ii. last year

EXAM PRACTICE

1. The scatter diagram shows the age of some cars and their values.

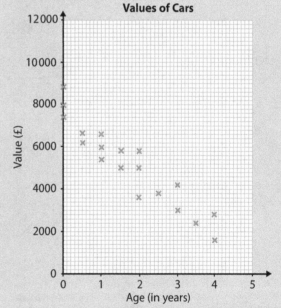

   a. Draw a line of best fit.   [1 mark]
   b. Use your line of best fit to estimate the age of a car when its value is £5000.   [2 marks]
   c. Use your line of best fit to estimate the value of a $3\frac{1}{2}$ year-old car.   [2 marks]
   d. Why can you not predict the value of a 10-year-old car using this diagram?   [1 mark]

Averages

Averages of Continuous Data

When data is grouped into **class intervals**, the exact data is not known.

You estimate the **mean** by using the midpoints of the class intervals.

> Add in two extra columns – one for the midpoint and one for fx.

Weight (W kg)	Frequency (f)	Midpoint (x)	fx
$30 \leqslant W < 35$	6	32.5	195
$35 \leqslant W < 40$	14	37.5	525
$40 \leqslant W < 45$	22	42.5	935
$45 \leqslant W < 50$	18	47.5	855
	60		**2510**

Σf Σfx

For continuous data:

$$\bar{x} = \frac{\Sigma fx}{\Sigma f}$$

Σ means the sum of
f represents the frequency
$\bar{x}$ represents the mean
x represents the midpoint of the class interval

$$\bar{x} = \frac{\Sigma fx}{\Sigma f}$$

$$\bar{x} = \frac{2510}{60}$$

$$\bar{x} = \mathbf{41.8\dot{3}}$$

Modal class is $40 \leqslant W < 45$

This class interval has the highest frequency.

To find the class interval containing the **median**, first find the position of the median:

$$= \frac{\Sigma f + 1}{2}$$

$$= \frac{60 + 1}{2} = 30.5$$

The median lies halfway between the 30th and 31st values. The 30th and 31st values are in the class interval $40 \leqslant W < 45$.

Hence, the class interval in which the median lies is $\mathbf{40 \leqslant W < 45}$.

Stem and Leaf Diagrams

Stem and leaf diagrams are useful for recording and displaying information. They can also be used to find the mode, median and range of a set of data.

These are the marks gained by some students in a Maths test:

52	45	63	67
75	57	68	67
60	59	67	

In an ordered stem and leaf diagram, it would look like this:

4	5
5	2 7 9
6	0 (3) 7 7 7 8
7	5

Median

Key $6 \mid 3 = 63$ marks

The median is the sixth value = **63 marks**

The mode is **67 marks**.

The range (highest value – lowest value) is $75 - 45 = \mathbf{30 \ marks}$

If the same students sat a second Maths test, their results could be put into a **back-to-back stem and leaf diagram**. These are very useful when comparing two sets of data.

	Test 2			Test 1				
9	6	2	4	5				
9 7 5	1	1	5	2	7	9		
	3	1	0	6	0	3	7	7	7	8
			7	5				

Key   Test 1 $5|2 = 52$ marks
          Test 2 $1|5 = 51$ marks

Comparing the data in the back-to-back stem and leaf diagram, you can say that:

'In test 2, the median score of 55 marks is lower than the median score in test 1 of 63 marks. The range of the scores in test 2 is 21 marks, which is lower than the range of the scores in test 1 of 30 marks. So on average, the students did better in test 1 but their scores were more variable than in test 2.'

SUMMARY

- **For continuous data:**

  – **Estimate the mean using:**

  $$\bar{x} = \frac{\Sigma fx}{\Sigma f}$$

  Σ **means the sum of**
  f **represents the frequency**
  $\bar{x}$ **represents the mean**
  x **represents the midpoint of the class interval**

  – **Modal class is the class interval with the highest frequency.**

- **Use stem and leaf diagrams to record and display information. Do not forget to order the leaves and write a key.**

- **Back-to-back stem and leaf diagrams can be used to compare two sets of data.**

QUESTIONS

QUICK TEST

1. The heights, h cm, of some students are shown in the table.

Height (h cm)	Frequency	Midpoint	fx
$140 \leqslant h < 145$	4		
$145 \leqslant h < 150$	9		
$150 \leqslant h < 155$	15		
$155 \leqslant h < 160$	6		

Calculate an estimate for the mean of this data.

2. **a.** Draw an accurate stem and leaf diagram of this data.

27	28	36	42	50	18
25	31	39	25	49	31
33	27	37	25	47	40
7	31	26	36	9	42

   b. What is the median of this data?

EXAM PRACTICE

1. The table shows information about the number of hours that 50 children watched television for last week.

    Work out an estimate for the mean number of hours the children watched television.
    [4 marks]

Number of hours (h)	Frequency
$0 \leqslant h < 2$	3
$2 \leqslant h < 4$	6
$4 \leqslant h < 6$	22
$6 \leqslant h < 8$	13
$8 \leqslant h < 10$	6

Cumulative Frequency Graphs

With a **cumulative frequency graph**, it is possible to estimate the **median** of grouped data and the **interquartile range**.

Example

The cumulative frequency table opposite shows the marks of 94 students in a Maths exam.

a. Complete the cumulative frequency table for this data.

b. Draw a cumulative frequency graph for this data.

To do this we must plot the upper boundary of each class interval on the x-axis and the cumulative frequency on the y-axis.
Plot (20, 2) (30, 8) (40, 18) …
Join the points with a smooth curve.
Since no students had less than zero marks, the graph starts at (0, 0).

c. Use the cumulative frequency graph to find the median and interquartile range of the data.

> The cumulative frequency is a running total of all the frequencies.

Mark (m)	Frequency	Mark	Cumulative frequency
$0 < m \leqslant 20$	2	$\leqslant 20$	2 (2)
$20 < m \leqslant 30$	6	$\leqslant 30$	8 (2 + 6)
$30 < m \leqslant 40$	10	$\leqslant 40$	18 (8 + 10)
$40 < m \leqslant 50$	17	$\leqslant 50$	35 (18 + 17)
$50 < m \leqslant 60$	24	$\leqslant 60$	59 (35 + 24)
$60 < m \leqslant 70$	17	$\leqslant 70$	76 (59 + 17)
$70 < m \leqslant 80$	11	$\leqslant 80$	87 (76 + 11)
$80 < m \leqslant 90$	4	$\leqslant 90$	91 (87 + 4)
$90 < m \leqslant 100$	3	$\leqslant 100$	94 (91 + 3)

> This means that 94 students had a score of 100 or fewer.

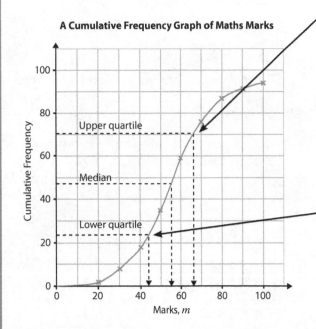

A Cumulative Frequency Graph of Maths Marks

The **upper quartile** is three-quarters of the way into the distribution:
$$\frac{3}{4} \times 94 = 70.5$$
Read across from 70.5 and down to the horizontal axis.
Upper quartile $\approx$ 67 marks

The **median** splits the data into two halves – the lower 50% and the upper 50%.
$$\text{Median} = \frac{1}{2} \times \text{cumulative frequency}$$
$$= \frac{1}{2} \times 94 = 47$$
Read across from 47. Median $\approx$ **56 marks**

The **lower quartile** is the value one-quarter of the way into the distribution:
$$\frac{1}{4} \times 94 = 23.5$$
Read across from 23.5. Lower quartile $\approx$ 44 marks

Interquartile range
= upper quartile – lower quartile
= 67 – 44 = **23 marks**

A large interquartile range indicates that the 'middle half' of the data is widely spread about the median.

A small interquartile range indicates that the 'middle half' of the data is concentrated about the median.

Box Plots

- **Box plots** are sometimes known as box and whisker diagrams.

- Cumulative frequency graphs are not easy to compare; a box plot shows the interquartile range as a box and the highest and lowest values as whiskers. Comparing the spread of data is then easier.

Example
The box plot of the cumulative frequency graph opposite would look like this:

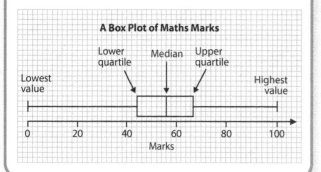

- **To draw a cumulative frequency graph, plot the upper boundary of each class interval on the x-axis and the cumulative frequency on the y-axis.**

- **Interquartile = upper quartile – range   lower quartile**

- **Use box plots to compare data.**

QUESTIONS

QUICK TEST

1. The box plot below shows the times in minutes to finish an assault course.

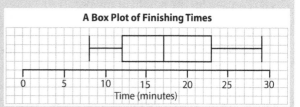

Use the box plot to complete the table below.

	Time (minutes)
Median time	
Lower quartile	
Interquartile range	
Longest time	

EXAM PRACTICE

1. Students in 9A and 9B took the same test. Their results were used to draw the following box plots.

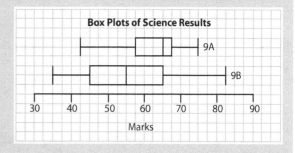

a. In which class was the student who scored the highest mark? [1 mark]

b. In which class did the students perform better in the test? You must give a reason for your answer. [3 marks]

Histograms

In a **histogram** the area of a bar is proportional to the frequency.

- If the class intervals have **equal widths**, frequency can be used for the height of the bar.

- If the class intervals have **unequal widths**, the height of the bar is adjusted by using **frequency density** and the area of the bar is equal to the frequency.

$$\text{Frequency density} = \frac{\text{frequency}}{\text{class width}}$$

The **modal class** is the class interval with the largest bar.

Drawing Histograms

The table below shows the time in seconds it takes people to swim 100 metres.

Time, t (seconds)	Frequency
$100 < t \leqslant 110$	2
$110 < t \leqslant 140$	24
$140 < t \leqslant 160$	42
$160 < t \leqslant 200$	50
$200 < t \leqslant 220$	24
$220 < t \leqslant 300$	20

- To draw a histogram where the class intervals are of different widths, you need to calculate the frequency densities. Add an extra column to the table.

The table should now look like this:

Time, t (seconds)	Frequency	Frequency density
$100 < t \leqslant 110$	2	$2 \div 10 = 0.2$
$110 < t \leqslant 140$	24	$24 \div 30 = 0.8$
$140 < t \leqslant 160$	42	$42 \div 20 = 2.1$
$160 < t \leqslant 200$	50	$50 \div 40 = 1.25$
$200 < t \leqslant 220$	24	$24 \div 20 = 1.2$
$220 < t \leqslant 300$	20	$20 \div 80 = 0.25$

- Draw on graph paper – make sure there are no gaps between the bars.

The histogram should look like this:

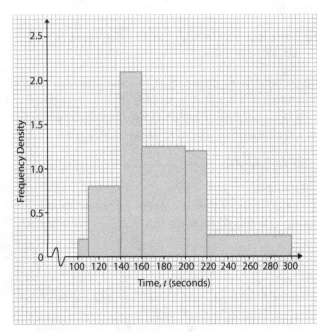

- Sometimes you will be asked to read from a histogram. In this case rearrange the formula for frequency density.

$$\text{Frequency} = \text{frequency density} \times \text{class width}$$

SUMMARY

- To draw a histogram with unequal class widths, you need to calculate the frequency densities:
  $$\text{Frequency density} = \frac{\text{frequency}}{\text{class width}}$$
  The areas of the bars are then equal to the frequencies they represent.

- To read from a histogram, you need to find the frequency:
  $$\text{Frequency} = \text{frequency density} \times \text{class width}$$

QUICK TEST

1. The table opposite gives some information about the ages of participants in a charity walk.

   On graph paper draw a histogram to represent this information.

Age (x) in years	Frequency
$0 < x \leqslant 15$	30
$15 < x \leqslant 25$	46
$25 < x \leqslant 40$	45
$40 < x \leqslant 60$	25

EXAM PRACTICE

1. The table and histogram give information about the distance (d km) travelled to work by some employees.

Distance (km)	Frequency	Frequency density
$0 < d \leqslant 15$	12	
$15 < d \leqslant 25$		
$25 < d \leqslant 30$	36	
$30 < d \leqslant 45$		
$45 < d \leqslant 55$	20	

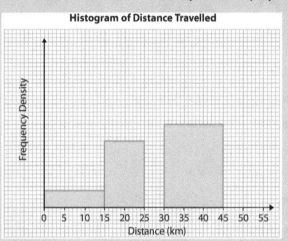

Histogram of Distance Travelled

   a. Use the information in the histogram to complete the table.   [2 marks]

   b. Use the table to complete the histogram.   [2 marks]

2. The histogram shows the masses of onions (in grams) in a sack.

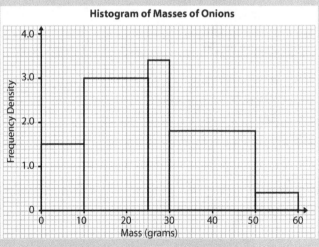

Histogram of Masses of Onions

   a. How many onions have a mass of between 10 and 25 grams?   [2 marks]

   b. How many onions were there in total?   [3 marks]

You are encouraged to show your working out, as you may be awarded marks for method in your exam even if your final answer is wrong. Full marks can be awarded where a correct answer is given without working being shown, but if a question asks for working out, you must show it to gain full marks. If you use a correct method that is not shown in the mark scheme below, you would still gain full credit for it.

N.B. In the answers to Exam Practice Questions, **[1]** indicates where individual marks may be awarded for correct workings and method.

Day 1

pages 4–5
Prime Factors, HCF and LCM
QUICK TEST

1. **a.** $50 = 2 \times 5 \times 5$ **or** 2×5^2

   b. $360 = 2 \times 2 \times 2 \times 3 \times 3 \times 5$ **or** $2^3 \times 3^2 \times 5$

   c. $16 = 2 \times 2 \times 2 \times 2$ **or** 2^4

2. **a.** False

   b. True

   c. True

   d. False

EXAM PRACTICE

1. $120 = \boxed{2} \times 2 \times 2 \times \boxed{3} \times 5$ **[1]**

   $42 = \boxed{2} \qquad \times \boxed{3} \quad \times 7$ **[1]**

   HCF = 6 **[1]**

2. LCM of 20 and 14

   $20 = 2 \times 2 \times 5$

   $14 = 2 \qquad \times 7$

   $LCM = 2 \times 2 \times 5 \times 7$ **[1]**

   $LCM = 140$

   Both buses will leave at the same time 140 minutes later, i.e. 2 hours and 20 minutes. **[1]**

   Time they leave together = 12.20pm **[1]**

You could also list both the times of the buses from Hatfield and St Albans and find the time that is the same in both lists.

pages 6–7
Fractions and Recurring Decimals
QUICK TEST

1. **a.** $\frac{13}{15}$

   b. $2\frac{11}{21}$

   c. $\frac{10}{63}$

   d. $\frac{81}{242}$

2. **a.** $\frac{7}{9}$

   b. $\frac{215}{999}$

   c. $\frac{16}{45}$

EXAM PRACTICE

1. Let x = number of pages in the magazine

   $\frac{3}{7}x = 12$

   $x = (12 \times 7) \div 3$ **[1]**

   $x = 28$ pages **[1]**

2. $x = 0.363\,636\ldots$ ①

   $100x = 36.363\,636\ldots$ ② **[1]**

   ② − ① $99x = 36$ **[1]**

   $x = \frac{36}{99}$

   $x = \frac{4}{11}$ **[1]**

pages 8–9
Indices
QUICK TEST

1. **a.** 6^8

   b. 12^{13}

   c. 5^6

   d. 4^2

2. **a.** $6b^{10}$

   b. $2b^{-16}$

c. $9b^8$

d. $\dfrac{1}{25x^4y^6}$ or $\dfrac{1}{25}x^{-4}y^{-6}$

EXAM PRACTICE

1. a. 1

b. $\dfrac{1}{7^2} = \dfrac{1}{49}$

c. $\sqrt[3]{64} \times \sqrt{144}$ **[1]**

$= 4 \times 12$

$= 48$ **[1]**

d. $\dfrac{1}{27^{\frac{2}{3}}} = \dfrac{1}{(\sqrt[3]{27})^2}$ **[1]**

$= \dfrac{1}{9}$ **[1]**

2. a. $\dfrac{x^{11}}{x^{15}} = x^{-4}$ **[1]**

$= \dfrac{1}{x^4}$ **[1]**

b. $\dfrac{12x^6}{2x^3}$ **[1]**

$= 6x^3$ **[1]**

pages 10–11
Standard Index Form
QUICK TEST

1. a. 6.4×10^4

b. 4.6×10^{-4}

2. a. 1.2×10^{11}

b. 2×10^{-1}

3. a. 1.4375×10^{18}

b. 5.476×10^{19}

EXAM PRACTICE

1. a. 4×10^7

b. 0.00006

2. $(2 \times 10^{-23}) \times (7 \times 10^{16})$ **[1]**

$= 14 \times 10^{-7}$ **[1]**

$= 1.4 \times 10^{-6}\,\text{g}$ **[1]**

Make sure that you check that your final answer is written in standard form.

pages 12–13
Surds
QUICK TEST

1. a. $2\sqrt{6}$

b. $10\sqrt{2}$

c. $6\sqrt{3}$

2. Molly is incorrect since
$(2 - \sqrt{3})^2 = (2 - \sqrt{3})(2 - \sqrt{3})$

$= 4 - 4\sqrt{3} + 3$

$= 7 - 4\sqrt{3}$

3. Correct since $\dfrac{1}{\sqrt{2}}$ has been rationalised to give $\dfrac{\sqrt{2}}{2}$

EXAM PRACTICE

1. $\dfrac{5 - \sqrt{75}}{\sqrt{3}} \times \dfrac{\sqrt{3}}{\sqrt{3}}$ **[1]**

$= \dfrac{5\sqrt{3} - \sqrt{225}}{3}$ **[1]**

$= \dfrac{5\sqrt{3}}{3} - 5$

$= -5 + \dfrac{5\sqrt{3}}{3}$

$\therefore a = -5$ **[1]**, $b = \dfrac{5}{3}$ **[1]**

2. $\dfrac{3}{\sqrt{6}} \times \dfrac{\sqrt{6}}{\sqrt{6}}$ **[1]**

$= \dfrac{3\sqrt{6}}{6} = \dfrac{\sqrt{6}}{2}$ **[1]**

3. $\dfrac{(5 + \sqrt{5})(2 - 2\sqrt{5})}{\sqrt{45}}$

$= \dfrac{10 - 10\sqrt{5} + 2\sqrt{5} - 2(\sqrt{5})^2}{\sqrt{45}}$ **[1]**

$= \dfrac{10 - 8\sqrt{5} - 10}{\sqrt{45}} = \dfrac{-8\sqrt{5}}{3\sqrt{5}}$ **[1]**

$= -\dfrac{8}{3}$ **or** $-2\dfrac{2}{3}$ **[1]**

pages 14–15
Upper and Lower Bounds
QUICK TEST

1. A = upper bound; E = lower bound

2. Upper bound = 0.226 (3 s.f.)

Lower bound = $0.2\overset{\cdot\cdot}{1}8$

EXAM PRACTICE

1. $a = 7.65$ $b = 4.3$

Upper bound   Upper bound
= 7.655   = 4.35 **[1]**

Lower bound   Lower bound
= 7.645   = 4.25

Upper bound of $R = \dfrac{7.655 - 4.25}{4.25}$

= 0.801 176 47... **[1]**

Lower bound of $R = \dfrac{7.645 - 4.35}{4.35}$

= 0.757 471 264... **[1]**

Difference = 0.801 176 47... − 0.757 471 264...

= 0.0437 (3 s.f.) **[1]**

> *Make sure that you do not round until the end of the question.*

pages 16–17
Formulae and Expressions

QUICK TEST

1. a. $8a - 7b$

   b. $4a^2 - 8b^2$

   c. $2xy + 2xy^2$

2. a. $-\dfrac{31}{5} = -6\dfrac{1}{5}$

   b. 4.36 or $4\dfrac{9}{25}$

   c. 9

3. $u = \pm \sqrt{v^2 - 2as}$

4. $p = \dfrac{-(t + vq)}{q - 1} = \dfrac{(t + vq)}{1 - q}$

EXAM PRACTICE

1. Josh is correct since $3x^2$ means $3 \times x^2$, this gives $3 \times 2^2 = 12$ **[2]**

2 a. $y = 4(x - 1)^2$

$y = 4(4 - 1)^2$ **[1]**

$y = 4 \times 9$

$y = 36$ **[1]**

   b. $y = 4(x - 1)^2$

$\dfrac{y}{4} = (x - 1)^2$ **[1]**

$\pm\sqrt{\dfrac{y}{4}} = (x - 1)$ **[1]**

$x = \pm\sqrt{\dfrac{y}{4}} + 1$ **[1]**

3. $\dfrac{p - 4y}{5y + q} = q$

$q(5y + q) = p - 4y$ **[1]**

$5qy + q^2 = p - 4y$

$5qy + 4y = p - q^2$ **[1]**

$y(5q + 4) = p - q^2$ **[1]**

$y = \dfrac{p - q^2}{5q + 4}$ **or** $y = \dfrac{q^2 - p}{-5q - 4}$ **[1]**

Day 2

pages 18–19
Brackets and Factorisation

QUICK TEST

1. a. $x^2 + x - 6$

   b. $4x^2 - 12x$

   c. $x^2 - 6x + 9$

2. a. $6x(2y - x)$

   b. $3ab(a + 2b)$

   c. $(x + 2)(x + 2) = (x + 2)^2$

   d. $(x + 1)(x - 5)$

   e. $(x - 10)(x + 10)$

EXAM PRACTICE

1. a. $3t^2 - 4t$

   b. $4(2x - 1) - 2(x - 4)$

$= 8x - 4 - 2x + 8$ **[1]**

$= 6x + 4$ **[1]**

2. a. $y(y + 1)$

   b. $5pq(p - 2q)$
     [1 for 5pq; 1 for (p − 2q)]

   c. $(a + b)(a + b + 4)$

   d. $(x - 2)(x - 3)$ **[1 for identifying 2 and 3 as factors]**

3. $(a + b)^2 - 2b(a + b)$

$= (a + b)(a + b) - 2ab - 2b^2$

$= a^2 + 2ab + b^2 - 2ab - 2b^2$ **[1]**

$= a^2 - b^2$ **[1]**

$= (a - b)(a + b)$ **[1]**

> *Always check that the final algebraic expression is in its simplest form – if not, then simplify by factorising.*

pages 20–21
Equations 1
QUICK TEST

1. $x = 8$ **2.** $x = -5$ **3.** $x = -\frac{1}{2}$

4. $x = -3.25$ **5.** $x = -\frac{1}{2}$ **6.** $x = 17$

EXAM PRACTICE

1. a. $5x - 3 = 9$

$5x = 9 + 3$ **[1]**

$5x = 12$

$x = \frac{12}{5}$ **[1]**

$x = 2.4$

b. $7x + 4 = 3x - 6$

$7x + 4 - 3x = -6$ **[1]**

$4x = -6 - 4$

$4x = -10$ **[1]**

$x = -\frac{10}{4}$

$x = -2.5$ **[1]**

c. $3(4y - 1) = 21$

$12y - 3 = 21$ **[1]**

$12y = 21 + 3$ **[1]**

$12y = 24$

$y = 2$ **[1]**

2. a. $5 - 2x = 3(x + 2)$

$5 - 2x = 3x + 6$ **[1]**

$5 = 3x + 6 + 2x$ **[1]**

$5 - 6 = 5x$

$x = -\frac{1}{5}$ **[1]**

b. $\frac{3x - 1}{3} = 4 + 2x$

$3x - 1 = 3(4 + 2x)$ **[1]**

$3x - 1 = 12 + 6x$

$-1 - 12 = 6x - 3x$ **[1]**

$-13 = 3x$

$x = -\frac{13}{3}$ **or** $x = -4\frac{1}{3}$ **[1]**

3. Joe forgot to multiply the 3 and the −6 together. When he multiplied out the bracket, it should say:

$3x - 18 = 42$

pages 22–23
Equations 2
QUICK TEST

1. $x = 5.5$ cm, shortest length is

$2x - 5 = 6$ cm

2. $k + 3 = \frac{1}{3}$ $\qquad$ $k = -\frac{8}{3}$

3. $4^{2k} = 4^3$ $\quad$ $2k = 3$ $\quad$ $k = \frac{3}{2}$

4. $2^{3k-1} = 2^6$ $\quad$ $3k - 1 = 6$ $\quad$ $k = \frac{7}{3}$

EXAM PRACTICE

1. $x + 30 + x + 50 + x + 10 + 2x$

$= 360$ **[1]**

$5x + 90 = 360$

$5x = 360 - 90$

$5x = 270$, so $x = 54$ **[1]**

Smallest angle $= 64°$ **[1]**

2. $2x + 2 + 3x + 1 + x + 6 + 2x + 3$
$= 3(x + 1 + 2x + 2 + x - 1)$ **[1]**

$8x + 12 = 3(4x + 2)$

$8x + 12 = 12x + 6$ **[1]**

$6 = 4x$

$x = 1.5$ cm **[1]**

Perimeter of the triangle
$= (4 \times 1.5 + 2)$

$= 8$ cm **[1]**

3. $16^{2k} = 64^{k+1}$

$(2^4)^{2k} = (2^6)^{k+1}$ **[1]**

$2^{8k} = 2^{6k+6}$ **[1]**

$8k = 6k + 6$ **[1]**

$2k = 6$

$k = 3$ **[1]**

pages 24–25

Quadratic and Cubic Equations

QUICK TEST

1. a. $(x - 2)(x - 4) = 0 \quad x = 2, x = 4$

b. $(x + 4)(x + 1) = 0 \quad x = -4, x = -1$

c. $(x + 2)(x - 6) = 0 \quad x = -2, x = 6$

2. a. $x = -5, x = 3$

b. $x = -\frac{1}{2}, x = -2$

EXAM PRACTICE

1. a. Shaded area
$= (2x + 5)(x + 3) - (x + 1)^2$ **[1]**

$= (2x^2 + 11x + 15) - (x^2 + 2x + 1)$

Area $= x^2 + 9x + 14$ **[1]**

Since the shaded area $= 45$ cm^2

$45 = x^2 + 9x + 14$ **[1]**

So $x^2 + 9x + 14 - 45 = 0$

$x^2 + 9x - 31 = 0$ **[1]**

b. i. $x = \dfrac{-b \pm \sqrt{b^2 - 4ac}}{2a}$

$x = \dfrac{-9 \pm \sqrt{9^2 - (4 \times 1 \times -31)}}{2 \times 1}$ **[1]**

$x = 2.66$ (3 s.f.) **[1]** or

$x = -11.7$ (3 s.f.) **[1]**

ii. Reject $x = -11.7$

Perimeter $3.66... \times 4 = 14.6$ cm (3 s.f.)

2. $x_{n+1} = \sqrt[3]{40 - x_n}$

Work out the solution to 5 decimal places.

$x_0 = 3$

$x_1 = \sqrt[3]{40 - 3} = 3.332\,221\,852...$ **[1]**

$x_2 = \sqrt[3]{40 - 3.332\,221\,852...} = 3.322\,218\,546...$

$x_3 = \sqrt[3]{40 - 3.322\,218\,546...} = 3.322\,520\,629...$

$x_4 = \sqrt[3]{40 - 3.322\,520\,629...} = 3.322\,511\,508...$

$x_5 = \sqrt[3]{40 - 3.322\,511\,508...} = 3.322\,511\,783...$ **[1]**

$x_6 = \sqrt[3]{40 - 3.322\,511\,783...} = 3.322\,511\,775...$

$x_7 = \sqrt[3]{40 - 3.322\,511\,775...} = 3.322\,511\,775...$

Solution $x = 3.322\,51$ (5 d.p.) **[1]**

When working out the solutions to iteration, it is important that you write down the solution to at least 8 decimal places.

pages 26–27

Completing the Square

QUICK TEST

1. $p = -2, q = 3$

2. $d = 3, e = -4$

EXAM PRACTICE

1. a. $x^2 + 10x + 5 = (x + 5)^2 - 25 + 5$ **[1]**

$= (x + 5)^2 - 20$

$a = 5$ **[1]** and $b = -20$ **[1]**

b. Minimum value of -20 occurs when
$x = -5$

2. a. $2x^2 + 8x + 3$

Divide each term in the equation by 2:

$2\left[x^2 + 4x + \frac{3}{2}\right]$ **[1]**

Complete the square to get in the form $a(x + b)^2 + c$:

$= 2\left[(x + 2)^2 + \frac{3}{2}\right]$ **[1]**

$= 2\left[(x^2 + 4x + 4) + \frac{3}{2} - 4\right]$

$= 2\left[(x + 2)^2 - \frac{5}{2}\right]$

Multiplying out by 2 gives:

$2(x + 2)^2 - 5$

$a = 2, b = 2, c = -5$ **[1]**

b. Turning point occurs at
$x = -2$ and $y = -5$, i.e. $(-2, -5)$

pages 28–29
Simultaneous Linear Equations
QUICK TEST

1. $a = 4, \ b = -4.5$

2. $x = 2, \ y = 4$

EXAM PRACTICE

1. $5a - 2b = 19$ ①

$3a + 4b = 1$ ②

$15a - 6b = 57$ **[1]** ③

$15a + 20b = 5$ ④

④ − ③

$26b = -52$ so $b = -2$ **[1]**

Substitute $b = -2$ into ①:

$5a - (2 \times -2) = 19$ **[1]**

$5a + 4 = 19$

$5a = 19 - 4$

$5a = 15$, so $a = 3$ **[1]**

$a = 3, b = -2$

2. $5x + 4y = 22$ **[1]** ①

$3x + 5y = 21$ ②

$15x + 12y = 66$ **[1]**

$15x + 25y = 105$

$13y = 39$, so $y = 3$ **[1]**

$5x + (4 \times 3) = 22$ **[1]**

$5x = 10$, so $x = 2$ **[1]**

$x = 2, y = 3$

Hats = £2 each and balloons = £3 each.

Always check that your solutions work by substituting them back into the second equation.

Day 3

pages 30–31
Algebraic Fractions
QUICK TEST

1. a. no

b. yes

c. yes

2. a. $\dfrac{5x - 1}{(x + 1)(x - 1)}$

b. $\dfrac{2aq}{3c}$

c. $\dfrac{40(a + b)}{3}$

EXAM PRACTICE

1. $\dfrac{8x + 16}{x^2 - 4} = \dfrac{8(x + 2)}{(x + 2)(x - 2)}$ **[1]**

$= \dfrac{8}{x - 2}$ **[1]**

2. $\dfrac{x^2(6 + x)}{x^2 - 36} = \dfrac{x^2(6 + x)}{(x + 6)(x - 6)}$ **[1]**

$= \dfrac{x^2}{x - 6}$ **[1]**

3. $\dfrac{2x^2 + 7x - 15}{x^2 + 3x - 10} = \dfrac{(2x - 3)(x + 5)}{(x + 5)(x - 2)}$ **[2]**

$= \dfrac{2x - 3}{x - 2}$ **[1]**

Answers

4. $\dfrac{3}{x} - \dfrac{4}{x+1} = 1$

$\dfrac{3(x+1) - 4x}{x(x+1)} = 1$ **[1]**

$3x + 3 - 4x = x^2 + x$ **[1]**

$x^2 + 2x - 3 = 0$ **[1]**

$(x - 1)(x + 3) = 0$

$x = 1$ **[1]** or $x = -3$ **[1]**

pages 32–33
Sequences
QUICK TEST

1. a. $4n + 1$

  b. $2 - n$

  c. $2n$

  d. $3n + 2$

  e. $5n - 1$

EXAM PRACTICE

1. 5 7 9 11

The difference between the terms = 2 **[1]**

$2 \times 1 = 2$ adjust by adding 3

nth term = $2n + 3$ **[1]**

2. $2n^2 + 1 = 101$

$2n^2 = 101 - 1$

$2n^2 = 100$ **[1]**, so $n^2 = 50$

Chloe is not correct. Since 50 is not a square number, 101 is not in the sequence. **[1]**

3.   7 22 47 82 127 …

First difference   15 25 35 45

Second difference 10 10 10 **[1]**

$10 \div 2 = 5$ **[1]**

nth term = $5n^2 + 2$ **[1]**

> *When the nth term has been worked out, check that the sequence works by substituting one of the terms back in.*

pages 34–35
Inequalities
QUICK TEST

1. a. $x < 2.2$

  b. $\dfrac{4}{3} \leqslant x < 3$

  c. $x > -\dfrac{9}{5}$

EXAM PRACTICE

1. $-2, -1, 0, 1, 2, 3, 4$

2. $4 + x > 7x - 8$

$12 > 6x$ **[1]**

$x < 2$ **[1]**

3.

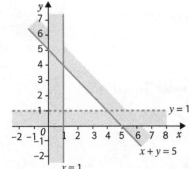

[1 for one correct line drawn; 1 for all three lines drawn; 1 for correct region shown]

4. $x^2 > 5x + 6$

$x^2 - 5x - 6 > 0$ **[1]**

$(x - 6)(x + 1) > 0$

$x < -1$ or $x > 6$ **[1 for critical values of –1 and 6; 1 for correct answer]**

pages 36–37
Straight-line Graphs
QUICK TEST

1. a.

x	−2	−1	0	1	2	3
y	−1	1	3	5	7	9

b.

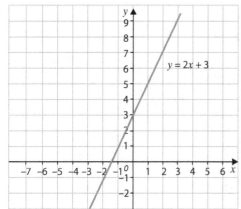

1. a. Gradient of EF =

$\frac{\text{change in } y}{\text{change in } x} = \frac{(6 - -4)}{(-2 - 3)}$ **[1]**

Gradient of EF $= \frac{10}{-5} = -2$ **[1]**

Goes through G (0, 3)

Equation of L: $y = -2x + 3$ **[1]**

b. Midpoint of EF $\left(\frac{-2 + 3}{2}, \frac{6 - 4}{2}\right)$ **[1]**

Midpoint $= \left(\frac{1}{2}, 1\right)$ **[1]**

c. Gradient of L $= -2$

Gradient of perpendicular is $\frac{1}{2}$ **[1]**

Equation of the line perpendicular to L:
$y = \frac{1}{2}x + 3$ **[1]**

pages 38–39
Curved Graphs
QUICK TEST

1. Graph A: $y = \frac{3}{x}$

Graph B: $y = 4x + 2$

Graph C: $y = x^3 - 5$

Graph D: $y = 2 - x^2$

EXAM PRACTICE

1. a.

x	−3	−2	−1	0	1	2	3
y	−28	−9	−2	−1	0	7	26

[2]

b.

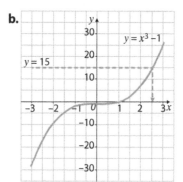

[3 if fully correct; 1 for correct labels; 1 for correct points joined by a curve]

c. $x = 2.5$

[1 for line at $y = 15$; 1 for $x = 2.5$]

d. 12 (approx.) **[1 mark for drawing tangent at $x = 2$; 1 mark for finding height/base of triangle; 1 mark for gradient = approx. 12]**

pages 40–41
Solving a Linear and a Non-linear Equation Simultaneously
QUICK TEST

1. a. $x = -1, y = 3$ $x = 2, y = 6$

 b. $x = 3, y = 4$ $x = -4, y = -3$

2. a. This is where the quadratic graph $y = x^2 + 2$ intersects the straight line graph $y = x + 4$. Their points of intersection are (−1, 3) and (2, 6).

 b. This is where the circle $x^2 + y^2 = 25$ intersects the straight line $y = x + 1$. Their points of intersection are (3, 4) and (−4, −3).

EXAM PRACTICE

1. $y - 5 = x$ ①

$x^2 + y^2 = 25$ ②

From ① $y = x + 5$ **[1]**

Substitute $y = x + 5$ into ②

$x^2 + (x + 5)^2 = 25$ **[1]**

$x^2 + (x + 5)(x + 5) = 25$

$x^2 + x^2 + 10x + 25 = 25$

$2x^2 + 10x + 25 - 25 = 0$

$2x^2 + 10x = 0$ **[1]**

$2x(x + 5) = 0$

Either $x = 0$ or $x = -5$ **[1]**

When $x = 0$, $y = 5$ and when $x = -5$, $y = 0$

Co-ordinates are (0, 5) **[1]** and (−5, 0) **[1]**

2. $x^2 + y^2 = 26$ ①

$y = 2x + 3$ ②

Substitute $y = 2x + 3$ into ①

$x^2 + (2x + 3)^2 = 26$ **[1]**

$x^2 + (2x + 3)(2x + 3) = 26$

$x^2 + 4x^2 + 12x + 9 = 26$ **[1]**

$5x^2 + 12x + 9 - 26 = 0$

$5x^2 + 12x - 17 = 0$ **[1]**

$(x - 1)(5x + 17) = 0$ **[1]**

Either $x = 1$ or $x = -3.4$

When $x = 1$, $y = 5$ **[1]** and when $x = -3.4$, $y = -3.8$ **[1]**

3. Gradient of tangent is $-\frac{3}{4}$ **[1]**

$4 = \left(-\frac{3}{4} \times 3\right) + c$ **[1]**

$c = \frac{25}{4}$

Equation of tangent is
$y = -\frac{3}{4}x + \frac{25}{4}$ **or** $4y + 3x = 25$ **[1]**

Day 4

pages 42–43
More Graphs and Other Functions

QUICK TEST

1. True for both solutions

2. False for both solutions

3. True for $x = -0.7$, false for $x = 5.6$

EXAM PRACTICE

1. Substituting into $v = ab^t$

$9000 = ab^0$ $4000 = ab^2$ **[1]**

Hence $a = 9000$, since $b^0 = 1$ **[1]**

$b = \sqrt{\dfrac{4000}{9000}}$

$b = \dfrac{2}{3}$ **[1]**

pages 44–45
Functions and Transformations

QUICK TEST

1. a.

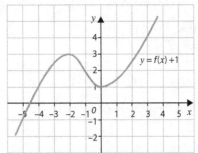

b.

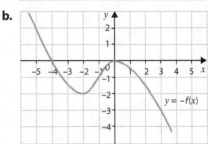

EXAM PRACTICE

1. **a.** (−1, −3)   **b.** (2, −7)

    c. (4, −3)   **d.** (−2, −3)

    e. (1, −3)

2. a. $fg(x) = 2x^2 + 3$ **[2] [1 for $2x^2$]**

b. $f^{-1}(x) = \frac{x-3}{2}$ **[2] [1 for $x - 3$]**

pages 46–47
Repeated Percentage Change
QUICK TEST

1. 51.7% (3 s.f.)

2. £121 856

EXAM PRACTICE

1. Savvy Saver $\frac{2.5}{100} \times 3000 = £75$

£75 × 2 = £150 **[1]**

Money Grows = $1.025^2 \times 3000 = £3151.88$

Interest earned = £151.88 **[1]**

Shamil is not correct – he would earn £1.88 more with the Money Grows investment. **[1]**

> *The use of the multiplier enables you to answer the question in an efficient manner.*

pages 48–49
Reverse Percentage Problems
QUICK TEST

1. a. £57.50

b. £127

c. £237.50

d. £437.50

EXAM PRACTICE

1. 100% – 12% = 88%

0.88 is the multiplier. **[1]**

$0.88x = 220$

$x = \frac{220}{0.88}$ **[1]** so $x = £250$ **[1]**

2. Yes, Joseph is correct **[1]** since $\frac{60}{0.85} = £70.59$ **[2]**

pages 50–51
Ratio and Proportion
QUICK TEST

1. £20 : £40 : £100

2. £35.28

3. 3 days

EXAM PRACTICE

1. There are many ways to work this out – this shows one possible way:
Work out the cost of 25 ml for each tube of toothpaste.
50 ml = £1.24; 25 ml = 62p
75 ml = £1.96; 25 ml = 65.3̇p
100 ml = £2.42; 25 ml = 60.5p **[2]**
The 100 ml tube of toothpaste is the better value for money. **[1]**

2. £52 × $1.49 = $77.48 **[1] or**

$63 ÷ 1.49 = £42.28

Cheaper in America by £9.72 (**or** by $14.48). **[1]**

pages 52–53
Proportionality
QUICK TEST

1. a. $y = kx$ **b.** $y = \frac{k}{\sqrt[3]{x}}$

c. $y = \frac{k}{x}$ **d.** $y = kx^3$

EXAM PRACTICE

1. $a = k\sqrt{x}$ **[1]**

$8 = k\sqrt{4}$

$\therefore k = 4$ **[1]**

$a = 4\sqrt{x}$ **[1]**

$64 = 4\sqrt{x}$ $\therefore x = 256$ **[1]**

2. a. $y = \frac{k}{x^2}$ **[1]**

$10 = \frac{k}{3^2}$ **[1]**

$k = 90$

$y = \frac{90}{x^2}$ **[1]**

$y = \frac{90}{4}$ so $y = 22.5$ **[1]**

b. $6 = \frac{90}{x^2}$

$x^2 = \frac{90}{6}$ **[1]**

$x = \sqrt{\frac{90}{6}}$ so $x = \pm 3.87$ **[1]**

> *On any proportionality question always work out the formula first.*

Answers

Day 5

pages 54–55
Similarity and Congruency 1
QUICK TEST

1. **a.** 20.9 cm   **b.** 13.8 cm   **c.** 4.5 cm

2. Yes they are congruent since $76° + 48° + 56° = 180°$. Hence both triangles have two sides the same and the included angle of 56° (SAS).

EXAM PRACTICE

1. $\dfrac{x}{12.4} = \dfrac{16.2}{9.7}$ **[1]**

   $x = \dfrac{16.2}{9.7} \times 12.4$ **[1]**

   $x = 20.7$ cm (3 s.f.) **[1]**

2. There are several ways of proving this. An example might be:

   Since triangle CDE is equilateral, length CD = length CE **[1]**. Both triangles have a common length, CF **[1]**, which is the perpendicular bisector of DE.

   Hence length DF = length EF.

   By SSS triangle CFD is congruent to triangle CFE. **[1]**

pages 56–57
Similarity and Congruency 2
QUICK TEST

1. 28.4 cm² (3 s.f.)

2. 78.75 cm²

3. 2.1875 cm³

EXAM PRACTICE

1. Length scale factor (k) is $8 : 24 = 1 : 3$ **[1]**

   Volume scale factor (k^3) = $1 : 27$ **[1]**

   Mass of larger solid is $27 \times 70 = 1890$ g **[1]**

2. Area scale factor (k^2) is $2160 : 60 = 36 : 1$ **[1]**

   Length scale factor (k) is $\sqrt{36} : 1 = 6 : 1$ **[1]**

   Volume scale factor (k^3) = $216 : 1$ **[1]**

   $270 \times 216 = 58\,320$ cm³ **[1]**

pages 58–59
Loci
QUICK TEST

1.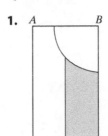

EXAM PRACTICE

1.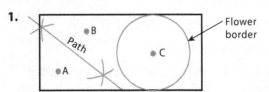

[1 for path; 1 for flower border]

pages 60–61
Angles
QUICK TEST

1. **a.** $a = 55°$

   b. $a = 74°, b = 32°$

   c. $a = 41°, b = 41°, c = 41°, d = 139°$

2. 30°

EXAM PRACTICE

1. For BE and CF to be parallel, the angles between the parallel lines are supplementary **[1]** and must add up to 180°.

   $53° + 127° = 180°$ **[1]**

2. Total sum of interior angles $= (6 - 2) \times 180°$

   $= 720°$ **[1]**

   $y + 126° + 83° + 145° + 138° + 79° = 720°$ **[1]**

   $y + 571° = 720°$

   $y = 720° - 571°$ **[1]**

   $y = 149°$ **[1]**

3. Exterior angle of the hexagon $= \dfrac{360°}{6}$

$\qquad\qquad = 60°$

Interior angle of the hexagon $= 180° - 60°$

$\qquad\qquad\qquad = 120°$ **[1]**

Exterior angle of the octagon $= \dfrac{360°}{8}$

$\qquad\qquad = 45°$

Interior angle of the octagon $= 180° - 45°$

$\qquad\qquad\qquad = 135°$ **[1]**

$x + 135° + 120° = 360°$

$\qquad\qquad x = 105°$ **[1]**

pages 62–63
Translations and Reflections
QUICK TEST

1. a. Reflection in $y = 0$ (x-axis)

  b. Reflection in $x = 0$ (y-axis)

  c. Translation of $\begin{pmatrix} -6 \\ 0 \end{pmatrix}$

  d. Translation of $\begin{pmatrix} 5 \\ -1 \end{pmatrix}$

EXAM PRACTICE

1. a. b. See below

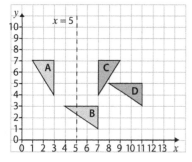

[2 for shape C; 1 for shape D]

  c. Reflection **[1]** in the line $y = x$ **[1]**

pages 64–65
Rotation and Enlargement
QUICK TEST

1.

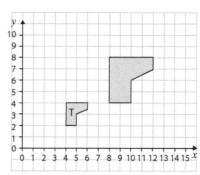

2.

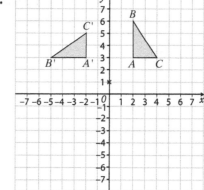

EXAM PRACTICE

1.

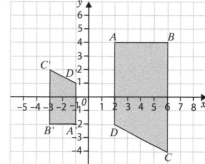

[1 for correct orientation; 1 for shape half the size; 1 for correct position]

pages 66–67
Circle Theorems
QUICK TEST

1. **a.** 62°
   b. 109°
   c. 53°
   d. 50°
   e. 126°

EXAM PRACTICE

1. **a.** 19° **[1]** A tangent and radius meet at 90° **[1]**
   b. 71° **[1]** Alternate Segment Theorem or angle in a semicircle is 90° **[1]**
   $E\hat{F}G = 90°$. Hence $G\hat{E}F = 71°$.

2.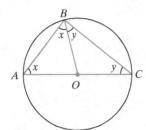

   $OA = OB = OC$, since they are all radii.

   Triangle OAB and triangle OBC are isosceles. **[1]**

   Each triangle contains a pair of equal angles and since angles in a triangle add up to 180°:

   $x + x + y + y = 180°$ **[1]**

   $2x + 2y = 180°$

   Hence $x + y = 90°$ **[1]**

pages 68–69
Pythagoras' Theorem
QUICK TEST

1. **a.** 17.46 cm (2 d.p.)
   b. 9.38 cm (2 d.p.)

EXAM PRACTICE

1. Since $26^2 = 24^2 + 10^2$ **[1]**

   $676 = 576 + 100$. Since this obeys Pythagoras' Theorem, then the triangle must be right-angled. **[1]**

2. Length of diagonal: $\sqrt{(5^2 + 3^2)} = 5.83...$

   Perimeter $= 5 + 4 + 7 + 5.83...$

   Perimeter $= 21.83...$ m **[1]**

   Number of lengths: $21.83... \div 2.5 = 8.732$

   9 lengths of beading needed $\times$ £1.74 **[1]**

   $=$ £15.66 **[1]**

   Always consider the context of the question. Here you round to 9 lengths of beading, since you would not be able to purchase 8.732 lengths.

3. $\sqrt{(17 - 7)^2 + (9 - 2)^2}$ **[1]**

   $= \sqrt{149}$ **[1]**

4. $\sqrt{5^2 + 5^2 + 5^2}$ **[1]**

   $= \sqrt{75}$ **[1]**

   $= 8.7$ cm (1 d.p.) **[1]**

Day 6

pages 70–71
Trigonometry
QUICK TEST

1. **a.** 5.79 cm
   b. 8.40 cm

2. **a.** 38.7°
   b. 43.0°

EXAM PRACTICE

1. **a.** $\tan PBN = \dfrac{\text{opp}}{\text{adj}}$

   $\tan PBN = \dfrac{19.7}{12.6}$ **[1]**

   $PBN = \tan^{-1}\left(\dfrac{19.7}{12.6}\right)$ **[1]**

   Angle $PBN = 57.4°$ (3 s.f.) **[1]**

   b. $\sin 48° = \dfrac{19.7}{PA}$ **[1]**

   $PA = \dfrac{19.7}{\sin 48°}$ **[1]**

   $PA = 26.5$ m (3 s.f.) **[1]**

pages 72–73

Trigonometry in 3D

QUICK TEST

1. a. i. True   **ii.** False

    iii. False   **iv.** True

  b. 10° (nearest degree)

EXAM PRACTICE

1. a. $AC = \sqrt{8^2 + 8^2}$

    $MC = \dfrac{\sqrt{128}}{2}$ **[1]**

    $EM^2 = 20^2 - \left(\dfrac{\sqrt{128}}{2}\right)^2$ **[1]**

    $EM = \sqrt{368}$ **[1]**

    $EM = 19.18$ cm (2 d.p.) **[1]**

  b. $\tan EDM = \dfrac{\text{opp}}{\text{adj}}$

    $\tan EDM = \dfrac{19.18...}{5.65...}$ **[1]**

    Angle $EDM = \tan^{-1}\left(\dfrac{19.18...}{5.65...}\right)$ **[1]**

       $= 73.6°$ (3 s.f.) **[1]**

pages 74–75

Sine and Cosine Rules

QUICK TEST

1. a. Correct   **b.** Correct

2. a. 114° (nearest degree)

  b. 53° (nearest degree)

EXAM PRACTICE

1. $\dfrac{AC}{\sin 32°} = \dfrac{16}{\sin 26°}$

    $AC = \dfrac{16}{\sin 26°} \times \sin 32°$ **[1]**

    $AC = 19.34...$ m **[1]**

  $\sin 58° = \dfrac{AB}{19.34...}$ **[1]**

    $AB = 19.34... \times \sin 58°$ **[1]**

    $AB = 16.4$ m (3 s.f.) **[1]**

> You could also use the sine rule to find AD and then use triangle ABD to get the correct answer.
>
> On more complex calculations in trigonometry, remember not to round off your answers until the end of the question.

pages 76–77

Arc, Sector and Segment

QUICK TEST

1. a. 14.05 cm   **b.** 49.17 cm²   **c.** 26.97 cm²

2. True since

$$\left(\frac{70°}{360°} \times \pi \times 8^2 - \frac{1}{2} \times 8 \times 8 \times \sin 70°\right) = 9.03 \text{ cm}^2 \text{ (3 s.f.)}$$

EXAM PRACTICE

1. Let angle $AOB = x$

    $\dfrac{x}{360°} \times 2\pi \times 6 = 10$ **[1]**

    $x = 95.49°$ **[1]**

    Area $= \dfrac{95.49°...}{360°} \times \pi \times 6^2$ **[1]**

    Area $= 30$ cm² (2 s.f.) **[1]**

pages 78–79

Surface Area and Volume

QUICK TEST

1. a. 565 cm³   **b.** 637 cm³

    c. 905 cm³   **d.** 245 cm³

EXAM PRACTICE

1. $\dfrac{1}{3}\pi r^2 h = 15$ **[1]**

    $r^2 = \dfrac{15 \times 3}{\pi \times 2.1}$ **[1]**

    $r = \sqrt{6.8209...}$ **[1]**

    $r = 2.61$ m (3 s.f.) **[1]**

2. $\pi \times (2x)^2 \times h = \dfrac{4}{3} \times \pi \times (3x)^3$ **[1]**

    $4x^2 \times \pi \times h = \dfrac{4}{3} \times \pi \times 27x^3$

    $4\pi x^2 h = 36\pi x^3$ **[1]**

    $h = \dfrac{36\pi x^3}{4\pi x^2}$ **[1]**

    $h = 9x$ **[1]**

pages 80–81

Vectors

QUICK TEST

1. a. i. $-\mathbf{a} + \mathbf{c}$ **ii.** $-\mathbf{c} - \mathbf{a}$

  b. $\mathbf{a} + \dfrac{1}{2}(-\mathbf{a} + \mathbf{c}) = \dfrac{1}{2}\mathbf{a} + \dfrac{1}{2}\mathbf{c}$

    or $\dfrac{1}{2}(\mathbf{a} + \mathbf{c})$

EXAM PRACTICE

1. a. i. $-\mathbf{a} + \mathbf{b}$ ii. $-2\mathbf{a} + 2\mathbf{b}$

   b. $-\mathbf{a} + \mathbf{b}$

   c. Since $\overrightarrow{FE} = -\mathbf{a} + \mathbf{b}$
   and $\overrightarrow{AD} = 2(-\mathbf{a} + \mathbf{b}) = 2\overrightarrow{FE}$ [1]

   So AD is twice the length of and parallel to FE. [1]

Day 7

pages 82–83
Probability
QUICK TEST

1. a.

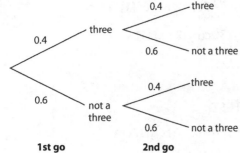

1st go   **2nd go**

   b. i. 0.16

      ii. 0.48

EXAM PRACTICE

1.

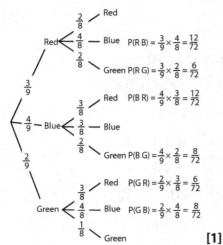

$$P(R\,B) = \frac{3}{9} \times \frac{4}{8} = \frac{12}{72}$$
$$P(R\,G) = \frac{3}{9} \times \frac{2}{8} = \frac{6}{72}$$
$$P(B\,R) = \frac{4}{9} \times \frac{3}{8} = \frac{12}{72}$$
$$P(B\,G) = \frac{4}{9} \times \frac{2}{8} = \frac{8}{72}$$
$$P(G\,R) = \frac{2}{9} \times \frac{3}{8} = \frac{6}{72}$$
$$P(G\,B) = \frac{2}{9} \times \frac{4}{8} = \frac{8}{72}$$

[1]

P(two different colours)

$$= \frac{12}{72} + \frac{6}{72} + \frac{12}{72} + \frac{8}{72} + \frac{6}{72} + \frac{8}{72} \,\text{[2]} = \frac{52}{72}$$

$$= \frac{13}{18} \,\text{[1]}$$

Or 1 − P(same colours)

$$= 1 - \left(\left(\frac{3}{9} \times \frac{2}{8}\right) + \left(\frac{4}{9} \times \frac{3}{8}\right) + \left(\frac{2}{9} \times \frac{1}{8}\right)\right) \text{[2]}$$

$$= 1 - \frac{20}{72} = \frac{52}{72}$$

$$= \frac{13}{18} \,\text{[1]}$$

2.

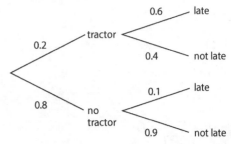

   a. P(meets a tractor and is late) $= 0.2 \times 0.6$ [1]
                                    $= 0.12$ [1]

   b. P(not late for work) $= (0.2 \times 0.4) + (0.8 \times 0.9)$ [1]
                           $= 0.08 + 0.72$ [1]
                           $= 0.8$ [1]

Although the question does not tell you to draw a tree diagram, doing so can be helpful.

pages 84–85
Sets and Venn Diagrams
QUICK TEST

1. a. $C \cup D = \{2, 3, 4, 5, 6, 7, 8, 9\}$

   b. $C \cap D = \{6, 7\}$

EXAM PRACTICE

1. a.

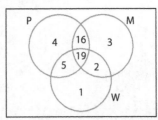

**[1 for 19 in the overlap of all three ovals;
1 for four of the seven values correct;
1 for a fully correct Venn diagram]**

   b. $4 + 16 + 19 + 5 = 44$ [1]

      P(plain chocolate) $= \frac{44}{50}$ [1] $= \frac{22}{25}$

   c. $16 + 5 = 21$ [1]

      P(likes one other type of chocolate) $= \frac{21}{44}$ [1]

pages 86–87
Statistical Diagrams
QUICK TEST

1. **a. i.** Roses   **ii.** Carnations

   b. i. 5%   **ii.** 15%

EXAM PRACTICE

1. **a.** Line of best fit should be as close as possible to all points and in the direction of the data.

   b. Approx. 2 years old **[2]**

   c. Approx. £3000 **[2]**

   d. You cannot predict the value of a 10-year-old car because this is beyond the range of the data and the relationship may not hold.

pages 88–89
Averages
QUICK TEST

1. 150.9

2. **a.**
```
0 | 7  9
1 | 8
2 | 5  5  5  6  7  7  8
3 | 1  1  1  3  6  6  7  9
4 | 0  2  2  7  9
5 | 0
```
       Key $4 | 2 = 42$

   b. Median = 31

EXAM PRACTICE

1. Mean $= \dfrac{\Sigma fx}{\Sigma f} = \dfrac{276}{50}$

   [1 for frequency × midpoint; 1 for totalling frequency; 1 for $\frac{276}{50}$]

   Mean = 5.52 hours **[1]**

> *Adding two extra columns to the table, i.e. midpoint (x) and fx, are helpful when working out the estimate of the mean.*

pages 90–91
Cumulative Frequency Graphs
QUICK TEST

1. Table completed as follows: Median time = 17 minutes; Lower quartile = 12 minutes; Interquartile range = 11 minutes; Longest time = 29 minutes

EXAM PRACTICE

1. **a.** 9B

   b. Class 9A **[1]**. The median is higher than 9B **[1]**: 50% of students in 9A scored over 65 marks. The top 50% of students in class 9B scored over 55 marks. **[1]**

pages 92–93
Histograms
QUICK TEST

1.
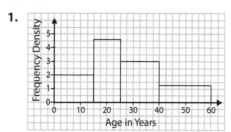

EXAM PRACTICE

1. **a.**

Distance (km)	Frequency	Frequency density
$0 < d \leqslant 15$	12	0.8
$15 < d \leqslant 25$	32 **[1]**	3.2
$25 < d \leqslant 30$	36	7.2
$30 < d \leqslant 45$	60 **[1]**	4.0
$45 < d \leqslant 55$	20	2.0

   b.
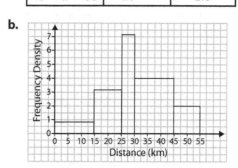

                                   [2]

2. **a.** frequency density × class width = frequency **[1]**

      $3 \times 15 = 45$ onions **[1]**

   b. $(1.5 \times 10) + (3 \times 15) + (3.4 \times 5) + (1.8 \times 20) +$ (0.4×10) **[2]** $= 117$ onions **[1]**

Glossary

Alternate angles – angles formed when two or more lines are cut by a transversal. If the lines are parallel then alternate angles are equal.

Alternate segment – the 'other' segment. In a circle divided by a chord, the alternate segment lies on the other side of the chord.

Arc – a curve forming part of the circumference of a circle.

Arithmetic sequence – a sequence with a common first difference between consecutive terms.

BIDMAS – an acronym that helps you remember the order of operations: Brackets, Indices and roots, Division and Multiplication, Addition and Subtraction.

Box plot – a graphic representation of the distribution of a set of data, showing the median, quartiles and the extremes of the data set.

Centre of enlargement – the point from which the enlargement happens.

Centre of rotation – the point around which a shape can rotate.

Chord – a line joining two points on the circumference of a circle.

Class interval – the width of a class or group, e.g. $0\,g < $ mass of spider $\leqslant 10\,g$

Coefficient – a number or letter multiplying an algebraic term.

Complete the square – a method of solving quadratic equations.

Compound interest – interest that accrues from the initial deposit plus the interest added on at the end of each year.

Conditional probability – the probability that an event will occur given that another event has already occurred.

Congruent – exactly alike in shape and size.

Constant of proportionality – the constant value of the ratio of two proportional quantities x and y.

Correlation – the relationship between the numerical values of two variables, e.g. there is a positive correlation between the numbers of shorts sold as temperature increases; there is a negative correlation between the age and the value of cars.

Corresponding angles – angles formed when a transversal cuts across two or more lines. When the lines are parallel, corresponding angles are equal.

Cumulative frequency – the running total of frequencies calculated from a frequency table.

Direct proportion – two values or measurements may vary in direct proportion, i.e. if one increases, then so does the other.

Empty set – a set containing no objects (members).

Enlargement – a transformation of a plane figure or solid object that increases the size of the figure or object by a scale factor but leaves it the same shape.

Equation – a number sentence where one side is equal to the other.

Expression – a statement that uses letters as well as numbers.

Exterior angle – an angle outside a polygon, formed when a side is extended.

Factorisation – finding one or more factors of a given number or algebraic expression.

Finite set – a set which has an exact number of members.

Formula – an equation that enables you to convert or find a value using other known values, e.g. area = length × width

Geometric sequence – a sequence with a common ratio.

Gradient – the measure of the steepness of a slope:
$$\frac{\text{vertical distance}}{\text{horizontal distance}}$$

Histogram – a chart that is used to show continuous data.

Identity – an identity is similar to an equation, but is true for all values of the variable(s); the identity symbol is $\equiv$ e.g. $2(x + 3) \equiv 2x + 6$

Independent events – two events are independent if the outcome of one event is not affected by the outcome of the other event, e.g. tossing a coin and throwing a dice.

Index (also known as power or exponent) – the small digit to the top right of a number that tells you the number of times a number is multiplied by itself, e.g. 5^4 is $5 \times 5 \times 5 \times 5$; the index is 4.

Inequality – a statement showing two quantities that are not equal.

Infinite set – a set which goes on and on, i.e. it has no end number.

Intercept – the point where a line or graph crosses an axis.

Interior angle – an angle between the sides inside a polygon.

Interquartile range – the difference between the lower quartile and the upper quartile, often found using a cumulative frequency graph.

Intersection – the point at which two or more lines cross.

Inverse (indirect) proportion – two quantities vary in inverse proportion when, as one quantity increases, the other decreases.

Irrational number – a number that cannot be written in the form $\frac{a}{b}$ where a and b are integers.

Iteration – the repetition of a process in order to find an approximate solution; the result of one iteration is used as the starting point for the next.

Locus (plural: loci) – the locus of a point is the path taken by the point following a rule or rules.

Lower bound – the bottom limit of a rounded number.

Lower quartile – the reading that is $\frac{1}{4}$ of the way up a cumulative frequency graph or a data set.

Mean – an average value found by dividing the sum of a set of values by the number of values.

Median – the middle item in an ordered sequence of items.

Midpoint – the point that divides a line into two equal parts.

Modal class – the largest class in a grouped frequency table.

Mode – the most frequently occurring value in a data set.

Multiplier – the number by which another number is multiplied.

Mutually exclusive events – two or more events that cannot happen at the same time, e.g. throwing a head and throwing a tail with the same toss of a coin are mutually exclusive events.

Percentage increase / decrease – the change in the proportion or rate per 100 parts.

Perpendicular bisector – a line drawn at right angles to the midpoint of a line.

Power – the small digit to the top right of a number that tells you the number of times a number is multiplied by itself, e.g. 5^4 is $5 \times 5 \times 5 \times 5$.

Prime factor – a factor that is also a prime number.

Prime number – a number with only two factors, itself and 1.

Pythagoras' Theorem – the theorem which states that the square on the hypotenuse of a right-angled triangle is equal to the sum of the squares on the other two sides.

Quadratic equation – an equation containing unknowns with maximum power 2, e.g. $y = 2x^2 - 4x + 3$. Quadratic equations can have 0, 1 or 2 solutions.

Quadratic graph – the U shaped graph of a quadratic equation.

Range – the spread of data; a single value equal to the difference between the greatest and the least values.

Ratio – the ratio of A to B shows the relative amounts of two or more things and is written without units in its simplest form or in unitary form, e.g. $A : B$ is $5 : 3$ or $A : B$ is $1 : 0.6$

Rational number – a number that can be written in the form $\frac{a}{b}$ where a and b are integers.

Reciprocal – the reciprocal of any number is 1 divided by the number (the effect of finding the reciprocal of a fraction is to turn it upside down), e.g. the reciprocal of $\frac{2}{3}$ is $\frac{3}{2}$

Reflection – a transformation of a shape to give a mirror image of the original.

Relative frequency – $\dfrac{\text{frequency of a particular outcome}}{\text{total number of trials}}$

Resultant – the result of adding two or more vectors together.

Roots – in a quadratic equation $ax^2 + bx + c = 0$, the roots are the solutions to the equation.

Rotation – a geometrical transformation in which every point on a figure is turned through the same angle about a given point.

Scalar – a quantity which has only magnitude.

Scale factor – the ratio by which a length or other measurement is increased or decreased.

Scalene – a triangle that has no equal sides or angles.

Sector – a section of a circle between two radii and an arc.

Set – a collection of objects (members).

Similar – the same shape but a different size.

Glossary

Simple interest – interest that accrues only from the initial deposit at the start of each year.

Simplify – making something easier to understand, e.g. simplifying an algebraic expression by collecting like terms.

Simultaneous equations – two or more equations that are true at the same time. On a graph the intersection of two lines or curves.

Standard index form (Standard form) – a shorthand way of writing very small or very large numbers; these are given in the form $a \times 10^n$, where a is a number between 1 and 10.

Stem and leaf diagram – a semi-graphical diagram used for displaying data by splitting the values.

Subset – a set within a set.

Substitution – to exchange or replace, e.g. in a formula.

Supplementary angles – angles that add up to 180°.

Surd – a number written as a square root, e.g. $\sqrt{3}$. A surd is an exact number.

Tangent – a straight line that touches a curve or the circumference of a circle at one point only.

Term – in an expression, any of the quantities connected to each other by an addition or subtraction sign; in a sequence, one of the numbers in the sequence.

Translation – a transformation in which all points of a plane figure are moved by the same amount and in the same direction. The movement can be described by a vector.

Tree diagram – a way of illustrating probabilities in diagram form. It has branches to show each event.

Trial and improvement – a method of solving an equation by making an educated guess and then refining it step-by-step to get a more accurate answer.

Trigonometry – the branch of mathematics that shows how to explain and calculate the relationships between the sides and angles of triangles.

Turning point – in a quadratic curve, a turning point is the point where the curve has zero gradient. It could be a minimum or a maximum point.

Universal set – contains all the objects being discussed.

Upper bound – the top limit of a rounded number.

Upper quartile – the reading that is $\frac{3}{4}$ of the way up a cumulative frequency graph or a data set.

Vector – a movement on the Cartesian plane described using a column, e.g. $\begin{pmatrix} 3 \\ 4 \end{pmatrix}$

ACKNOWLEDGEMENTS

The author and publisher are grateful to the copyright holders for permission to use quoted materials and images.

All images are © iStock/Thinkstock/Getty Images; © Shutterstock.com and © HarperCollinsPublishers Ltd

Every effort has been made to trace copyright holders and obtain their permission for the use of copyright material. The author and publisher will gladly receive information enabling them to rectify any error or omission in subsequent editions. All facts are correct at time of going to press.

Published by Collins

An imprint of HarperCollinsPublishers
1 London Bridge Street
London SE1 9GF

HarperCollinsPublishers
Macken House, 39/40 Mayor Street Upper,
Dublin 1, D01 C9W8, Ireland

ISBN: 9780008317676

Content first published 2016
This edition published 2020
Previously published as Letts
10 9 8 7 6

© HarperCollinsPublishers Limited 2020

British Library Cataloguing in Publication Data.
A CIP record of this book is available from the British Library.

Commissioning Editor: Emily Linnett
Author: Fiona Mapp
Project Management: Richard Toms
Cover Design: Kevin Robbins
Inside Concept Design: Ian Wrigley
Text Design and Layout: Q2A Media
Production: Lyndsey Rogers
Printed and Bound in the UK
by Ashford Colour Press Ltd

MIX
Paper | Supporting responsible forestry
FSC™ C007454

This book contains FSC™ certified paper and other controlled sources to ensure responsible forest management.

For more information visit: www.harpercollins.co.uk/green